Advances in African Economic, Social and Political Development

Series Editors

Diery Seck, CREPOL—Center for Research on Political Economy, Dakar Yoff, Senegal

Juliet U. Elu, Morehouse College, Atlanta, USA

Yaw Nyarko, New York University, New York, USA

Africa is emerging as a rapidly growing region, still facing major challenges, but with a potential for significant progress – a transformation that necessitates vigorous efforts in research and policy thinking. This book series focuses on three intricately related key aspects of modern-day Africa: economic, social and political development. Making use of recent theoretical and empirical advances, the series aims to provide fresh answers to Africa's development challenges. All the socio-political dimensions of today's Africa are incorporated as they unfold and new policy options are presented. The series aims to provide a broad and interactive forum of science at work for policymaking and to bring together African and international researchers and experts. The series welcomes monographs and contributed volumes for an academic and professional audience, as well as tightly edited conference proceedings. Relevant topics include, but are not limited to, economic policy and trade, regional integration, labor market policies, demographic development, social issues, political economy and political systems, and environmental and energy issues.

All titles in the series are peer-reviewed. The book series is indexed in SCOPUS.

Marieme Toure

Agricultural Adaptation to Climate Change in Sub-Saharan Africa

Evidence from Senegal

Springer

Marieme Toure
The New School for Social Research
New York, NY, USA

ISSN 2198-7262 ISSN 2198-7270 (electronic)
Advances in African Economic, Social and Political Development
ISBN 978-3-031-80715-2 ISBN 978-3-031-80716-9 (eBook)
https://doi.org/10.1007/978-3-031-80716-9

This Springer imprint is published by the registered company Springer Nature Switzerland AG
The registered company address is: Gewerbestrasse 11, 6330 Cham, Switzerland

Foreword by Nicoletta Batinia

Africa is at a pivotal juncture in its history, facing the twin challenges of climate change and agricultural sustainability. The impacts of a warming planet are all too evident in Senegal, where communities depend heavily on subsistence agriculture. The vulnerability of Africa to climate change is well documented. Rising temperatures, erratic rainfall patterns, and volatile extreme weather events wreak havoc upon agricultural productivity. For tens of millions of Africans, farming is not just a livelihood but a way of life, deeply intertwined with cultural traditions and community structures. Yet, as the population continues to grow, the pressure on agricultural systems intensifies, demanding innovative and sustainable solutions. The allure of industrial agriculture, with its promises of higher yields and economic growth, is tempting. However, this approach carries significant risks. Industrial agriculture is notoriously water-intensive and heavily reliant on fertilizers, chemicals, pesticides, herbicides, and genetically modified organisms. These practices can lead to severe environmental degradation, polluting water sources, depleting soil health, and ultimately undermining the resilience of agricultural systems. Children, the most vulnerable among us, stand to suffer the most if we fail to act. The future of Africa's youth is inextricably linked to the health of its environment and the sustainability of its agricultural practices. It is imperative that we pursue pathways to inclusive sustainability that honor and respect both Nature and the wisdom of local communities. There are several pathways to achieving this inclusive sustainability. These pathways must be rooted in a deep respect for Nature and a commitment to preserving the rich traditions and knowledge of local communities. By embracing agro ecological practices, promoting biodiversity, and supporting smallholder farmers, we can build resilient agricultural systems that not only feed the population but also protect the environment for future generations. This book aims to shed light on these critical issues and explore the innovative solutions that can help Senegal and the whole of Africa navigate the challenges of climate change and agricultural sustainability. It is a call to action for policymakers, researchers, and communities to work together in forging a sustainable future for Africa, one that ensures the well-being of its people and the preservation of its natural heritage. I highly recommend it.

Dr. Nicoletta Batinia
A Leading Member of the International
Monetary Fund Staff
Expert on Sustainable Agriculture
Senior Advisor at Sveriges Riksbank
The Central Bank of Sweden
Stockholm, Sweden

Foreword by Willi Semmler

I would enthusiastically support the publication of the book manuscript of Dr. Marieme Toure. It is a manuscript originally written as a Ph.D. dissertation at the New School for Social Research, New York, for which Marieme was awarded a distinctive prize in Spring 2024.

It is a manuscript on Sustainable Agriculture in Sub-Saharan Africa. This is an important topic regarding the issue of the current climate change and climate risks. As to this topic, the Sub-Saharan African countries and their agriculture are threatened by climate change, extreme weather events, declining agricultural productivity, and child malnutrition. This research focuses on those challenges as well as on how sustainable agriculture should be built up to respond to such enormous threats. Many of the Sub-Saharan countries are facing the fate of multiple climate vulnerabilities, deteriorating conditions for agricultural production, and malnutrition.

An important country in case is Senegal, where some new policies of sustainable agriculture are explored. Agricultural production and sustainable agriculture have serious implications to children's malnutrition. In the study, it is shown that climate variability and change stand to exacerbate malnutrition through effects on both the natural and social systems upon which humans depend, including food systems. An important further part of the study focuses on the financing of sustainable agriculture, ultimately, far-reaching actions need to be undertaken to achieve a green transition in Sub-Saharan African countries. There is a need to act now by combining strategies of mitigating climate risks, promotion of growth, sustainable agriculture, and child nutrition.

Prof. Willi Semmler
Arnhold Professor of International
Cooperation and Development
Department of Economics
The New School for Social Research
New York, USA

Acknowledgements

I am deeply grateful to my dissertation committee, and in particular, to my advisor Prof. Willi Semmler for all the help and guidance I have received from him throughout my doctoral studies. My conversations with Professor Semmler and his advanced macroeconomic classes and economics of climate change have substantially shaped my academic and career interests. Professor Nikolaos Chatzarakis was extremely helpful with his comments, and he acted as a co-advisor for this research. I am also thankful to Prof. Lucas Bernard and Dr. Nicoletta Batini for their to-the-point responses on my dissertation proposal and chapter drafts. I am also thankful to the Economics Department for their financial support with the NSSR Prize Fellowship from 2020 to 2023 academic years. Much appreciation to all my professors and colleagues at the Economics department.

This work could not have been done without the support of my family. They were always supportive and encouraging throughout my Ph.D. They did what they could to emotionally support me throughout this journey. Their presence and support were the main sources of my daily energy.

Executive Summary

Sub-Saharan African countries and their agriculture are threatened by climate change, extreme weather events, declining agricultural productivity, and child malnutrition. This research focuses on those challenges as well as on how sustainable agriculture should be built up responding to such enormous threats. Many of the Sub-Saharan countries are facing the fate of multiple climate vulnerabilities, deteriorating conditions for agricultural production, and malnutrition, an important country in case is Senegal, where some new policies of sustainable agriculture are explored. Highly vulnerable to shocks associated with climate change, Senegal's crop production has since long been subject to historical droughts and the recent return of rains and their flooding consequences. Since these climate-related weather extremes are expected to increase vulnerability to food security, particularly child malnutrition, this study is an effort to estimate the resulting impact of climate change on long-term crop production, food security, and child malnutrition. The visualization of annual maximum and minimum temperatures, and average annual rainfall from 1900 to 2020, shows that the country has warmed up so much that there is no turning back. The forecasting of the long-term cereal production of the four major staple crops in the country came out to be very vulnerable to not only the different scenarios of greenhouse gas emissions, but also to changes in technology and the quality of soil. These have serious implications to children's malnutrition, as shown later in the study. Climate variability and change stand to exacerbate malnutrition through effects on both the natural and social systems upon which humans depend, including food systems. An important further part of the study focuses on the financialization of agricultural productions by pointing out the importance of the parallel markets as well as the recent rush of overseas investors to Senegalese land taking. While the emissions of the West are responsible for the climate crisis and its consequences, Sub-Saharan Africa now has to pay for the climate damage and Senegal is no exception to this. Hence, the importance to develop sustainable and adaptation strategies such as: reforming agriculture in the direction of sustainable agriculture, matching private finance and green infrastructure and the creation of green finance in the form of green banks. Ultimately, far-reaching actions need to be undertaken to achieve a green transition in Sub-Saharan African countries. There is the need to act now by combining strategies of mitigating climate risks, promotion of growth, sustainable agriculture, and child nutrition.

Contents

List of Figures

List of Tables

Chapter 1
Climate Trends in Senegal

1.1 Introduction

Senegal is located in the western most part of Africa and bordered by Mauritania, Mali, Guinea, and Guinea-Bissau. The country has a population of 18.4 million with an annual population growth rate of 2.6% (World Bank, 2022a). The majority of the population lives on the Atlantic coast, especially in the Dakar area, which makes up only 0.3% of the country's territory, while hosting one-third of its population (World Bank, 2022b). With a real GDP per capita of 1,598.7 USD in 2022 (World Bank, 2022a), the country is a least developed country (LDC). Senegal's economy is dominated by the services sector, contributing 49.6% to the country's GDP in 2023, followed by the industrial sector with 24.67% and the agricultural sector with 15.32% (World Bank, 2020). Due to its coastline, the fisheries sector is one of the backbones of the economy, constituting an important source of employment and contributing around 12.5% to Senegal's total export value in 2021 (OOEC, 2021). Other key exports include gold, petroleum, phosphoric acid, and groundnuts, the latter of which is mostly exported to China (OOEC, 2021). Due to poor soil quality and erratic precipitation, the agricultural sector accounts for only 17.5% of the national GDP (World Bank, 2022b). However, the sector is the primary means of livelihood for Senegal's population, especially in rural areas. Important staple crops are millet, rice, cowpeas, maize, and sorghum while groundnuts, sugarcane, and cotton constitute major cash crops (FAO, 2015).

Senegal's topography is mostly flat and consists predominantly of rolling sandy plains but rises to hills in the south east. The lowest point of the country is on the Atlantic coast, while the highest point lies at 648 m close to the border with Guinea. The country has a mostly arid and semi-arid Sahelian climate defined by dry winds which originate in the continental interior and maritime winds which bring precipitation. Especially the north of the country is very dry, part of which can be ascribed to the north-eastern trade winds and the harmattan, a dry wind, which is strongest from late November to mid-March and carries large amounts of sand and dust (Huq et al.,

M. Toure, *Agricultural Adaptation to Climate Change in Sub-Saharan Africa*, Advances in African Economic, Social and Political Development,
https://doi.org/10.1007/978-3-031-80716-9_1

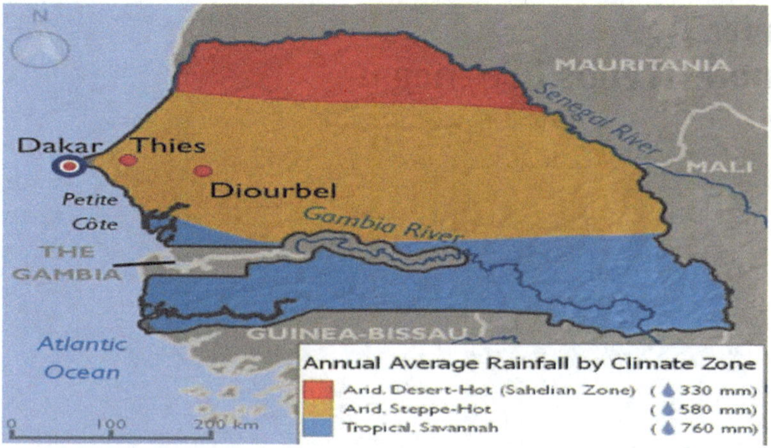

Fig. 1.1 Senegal climate classification. *Source* "Climate Risk Change Profile", by USAID (2020)

2006). The south of the country has higher amounts of precipitation. Accordingly, Senegal can be divided into three Agro-Ecological-Zones (AEZs): Arid/Sahel, Semi-arid/Sudan Savannah and Tropical Savannah (IIOTA, 2020). Each of these zones, as shown in Fig. 1.1, is characterized by specific temperature and moisture regimes, and consequently specific patterns of crop production and pastoral activities. While Senegal has abundant renewable water resources, surface and groundwater availability display a high seasonal and regional variability.

Senegal's climate is largely influenced by the proximity to the Atlantic Ocean coast and latitude. Mean annual temperatures range from 25 to 30 °C with lower values closer to the coast and higher values further inland. Annual precipitation sums range from 250 mm in northern Senegal, which has an arid desert climate, to 1450 mm in the south-western part of the country, which is characterized by a more tropical climate. Senegal has a single rainy season (unimodal precipitation regime) from April to November in the southern part of the country, with decreasing length and precipitation amounts toward the north (Stefan, 2016).

This chapter is organized into three sections. In Sect. 1.2, Senegal's historical temperature and rainfall trends from 1900 to present is discussed. After that, Sect. 1.3 focuses on the natural disaster events in the country, which is dominated by recurrent droughts and floods episodes. This chapter ends with Sect. 1.4, where we discuss the implications of climate change in food security and children's malnutrition.

1.2 Historical Temperature and Rainfall Trends from 1900 to Present

At latitudes of 3–12°N, most of Senegal has a typically tropical climate, but the northern regions lie in the sub-tropical semi-arid belt called the Sahel. The topography of Senegal consists mainly of rolling sandy plains, but rises to hills in the south east. Senegal has one wet season between July and September. There is a strong north-south gradient in total rainfall received at this time, with up to 200 mm per month in the south, but less than 100 mm per month in the very north. This rainfall season is controlled by the movement of the tropical rain belt (also known as the Inter-Tropical Conversion Zone, ITCZ) which oscillates between the northern and southern tropics over the course of a year, affecting Senegal when it is in its northern position (Mcsweeney et al., 2010).

Variations in the latitudinal movements of the ITCZ from one year to another cause large inter-annual, and inter-decadal, variability in this wet-season rainfall. The most well-documented cause of these variations is the El Niño Southern Oscillation (ENSO). El Niño events are associated with drier conditions in the Sahel. Temperatures in Senegal generally increase from the coast toward the west of Senegal in most seasons. The exception to this is in the wet season, July–August–September, when the cooling influence of cloud and rainfall mean that the south is cooler than the north. In the hottest season, April–May–June, the hottest regions have average temperatures of up to 35 °C, while the cooler coastal regions are 25–28 °C. In the cooler seasons (October–November–December and January–February–March) average temperatures can be below 25 °C at the coast and up to 30 °C in the west. Inter-annual variability in temperature in this region of Western Africa is caused by the El Niño Southern Oscillation (ENSO). In la Niña years, temperatures tend to be cooler than average throughout the year (Mcsweeney et al., 2010).

Rainfall reductions and temperature increases can be visualized in Fig. 1.2. Indeed, mean annual temperature has increased by 0.9 °C since 1960, an average rate of 0.20 °C per decade. Also, the rate of increase is most rapid in October–November–December, at 0.29 °C per decade. Sahelian rainfall is characterized by high variability on inter-annual and inter-decadal timescales, which can make long-term trends difficult to identify. A period of particularly high rainfall occurred in the early 1960s, while the early 80s were particularly dry. Furthermore, statistically significant decreases of around 10–15 mm per decade have, however, been observed in the southern regions of Senegal in the wet season (July–August–September) between 1960 and 2006. Some unusually high rainfalls have occurred in the dry season (January–February–March) in very recent years (2000–2006), but this has not been part of a consistent trend.

Moreover, rainfall in Senegal declined rapidly between 1950 and the mid-1980s, and partially recovered in the 1990s. Between 2000 and 2009, however, the recovery slowed, and the 2000–2009 mean remained about 15% lower than the 1920–1969 mean. The post-1980s rainfall increases are probably due to the warming of the

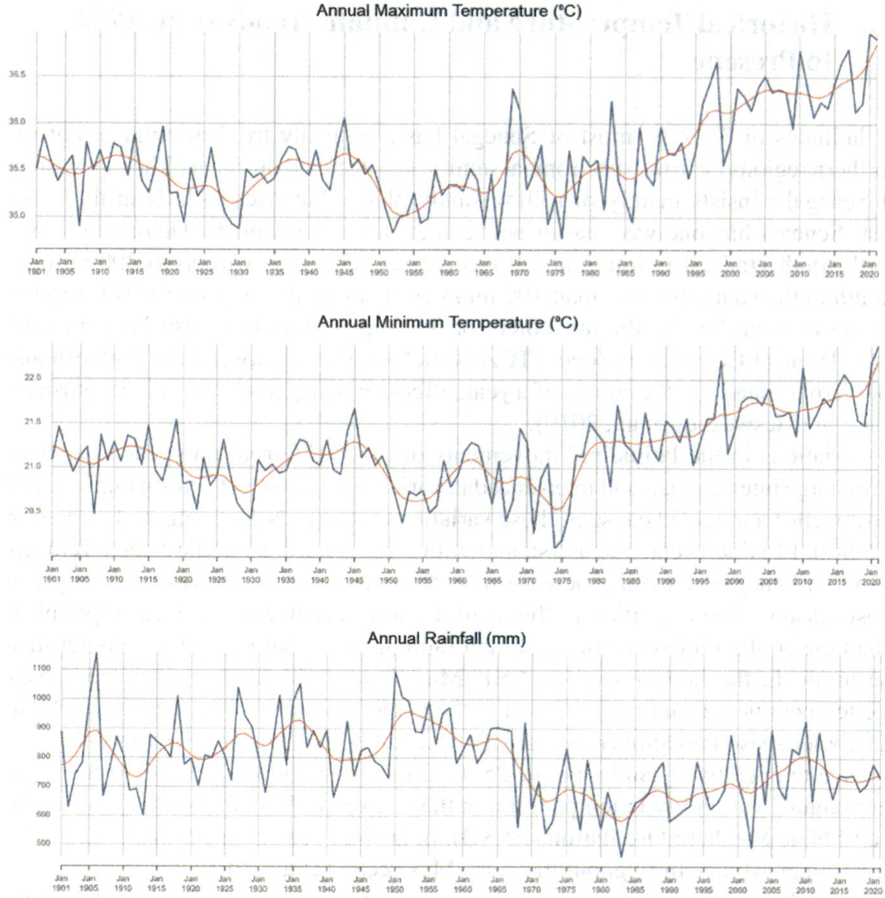

Fig. 1.2 Senegal historical climate trend. *Source* Author's computations based on Climate Research Unit (CRU)'s data

north Atlantic Ocean (Hoerling et al., 2006); as the northern tropical Atlantic has increased in temperature over this period, this has drawn the summer rains further north, increasing rainfall in the Sahel. These rainfall changes can be visualized in three ways: as a reduction of the areas receiving adequate rainfall for viable agricultural livelihoods, as maps of anticipated changes in rainfall, and as time series plots.

1.3 Historical Natural Disaster Events: From Droughts to Floods

Climate change is one of the most important issues on the global political and economic agenda, yet it has taken at least 20 years to become an international priority. In Africa, especially in Senegal, climate change is far from abstract—it is already determining the course of people's lives. Extreme weather events and greater unpredictability in weather patterns are having serious consequences for people who rely on land, lakes, and seas to feed themselves and to earn a living. Climate change presents an additional stress for Senegalese already struggling with the challenges posed by climate variability, ongoing environmental degradation, and widespread poverty.

Senegal is a land of geographic and climatic contrasts. Most of the country is subject to the weather conditions of the Sahel zone, which is characterized by a single rainy season that lasts for up to three months (Huq et al., 2006). The region is prone to unstable and irregular rainfalls, and has experienced recurrent droughts, for periods ranging from decades to centuries, that have, in some cases, been attributed to internal natural variability of the climate. The variability of rainfalls during the rainy season is increasing, making weather harder to predict and droughts more severe, particularly in northern Senegal, an area that is highly susceptible to desertification. Like its neighboring countries, Senegal was hit by a serious drought in the late 1960s, which has affected the country's ecology and environment.

1.3.1 Persistent Droughts Events from 1970s to 1980s

The period of drought in the Sahel began in the 1970s (Fig. 1.3). 1973 is embedded in the mentality of the Senegalese peasant as the year of the great drought and of the subsequent loss of crops, the extensive drying up of water sources, the difficulty in procuring water, the death of livestock, the food shortages in most households and, consequently, of migration. Since the 1970s, a cycle of drought has broken the economic and social equilibrium of the Senegalese rural regions and migration toward cities has become the most common adaptive solution (Fall, 1998).

Figure 1.2 illustrates that the irregularity of rainfall has a long history but became more pronounced from the 1970s. In the mid-1980s, a new extreme drought hit rural regions so dependent on rainfall and already impacted by new state liberalization policies. The long and intense periods of pluviometric deficit that marked the Sahelian region between 1970 and 1980 also affected Senegal, particularly in its northern regions. During that period, certain years were particularly catastrophic with respect to the limited rainfall: 1970 (653.6 mm) and 1972 (504.9 mm) because they were preceded by relatively rainy seasons (962.8 mm in 1969, 825 mm in 1971, 460 mm in 1977) (Fig. 1.2). The year 1983 (411 mm) showed even more of a deficit. Dagana

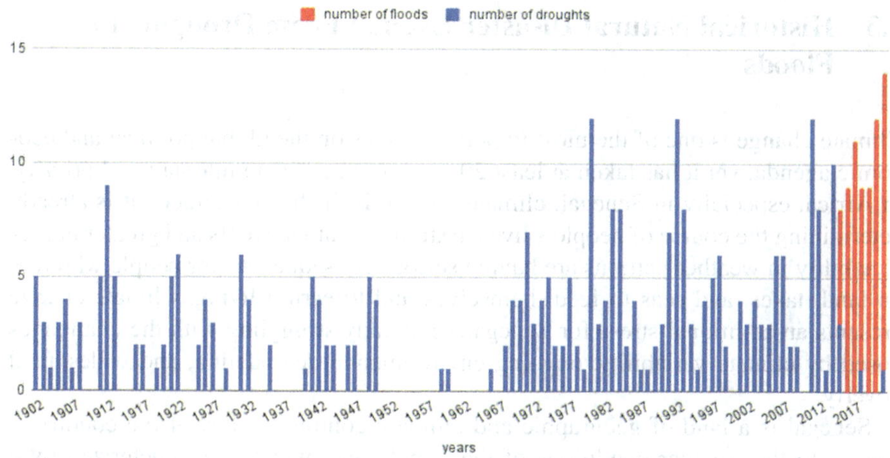

Fig. 1.3 Senegal historical natural disaster events: from droughts to floods. *Source* Author's computations based on U.S. Department of Commerce—NOAA (National Oceanic Atmospheric Administration) data

in the delta of the Senegal River (in the North) only saw 68 mm of rainfall during the entirety of 1983 (Gueye et al., 2015).

Gueye et al. (2015) noticed the shifting of the 400 mm isohyet from the latitude of Ferlo, a consistently dry area, to that of Kaolack in the center of the country, a region that constituted the capital of peanut production throughout the twentieth century. This 150–200 km shift to the south in less than two decades is dramatic. It obliged hundreds of thousands of rural citizens to adapt their calendar of activities and their cultural practices to the rigor of climatic conditions. The shifting of the 1000 mm isohyet from the latitude of Gambia to that of Ziguinchor, Sedhiou (in the South) or Kedougou (in the east) is also an important factor. It reveals a lesser-known type of 'drought' that is almost as destabilizing for populations known for their technical nature and their dependence on the cultivation of rice.

One of the direct consequences of the drought is the reduction of cultivated land due to an insufficient replenishment of seed funds as is the case with peanut crops, or due to the loss of fields that have become uncultivable because of wind erosion. According to the PANAUDIT cabinet, 'the total surface of cultivated land annually is 2.29 million hectares in Senegal, that is, about 60% of arable land.' Furthermore, according to the same study, 'the quantity of cultivated surfaces diminished by 4% between 1970–1980 and 1980–1994.' With the crisis in the all-important peanut crop, agriculture has all but broken down. The breakdown is linked not only to diminished precipitation but also to the reduced fertility of the land and a corresponding reduction in harvests that neither the weak agricultural mechanism nor the colonization of new spaces toward the eastern region of the country has been able to counter. The only perceived solution to this decrease in harvests was an increase in the total area of cultivated land. The greater amount of land cultivated for niébé and manioc crops due to the resistance of these crops to drought conditions or to their short vegetative cycles has not changed the overall situation significantly.

What is more, the exceptional Sahelian drought that occurred between the late 1960s and the mid-1980s highlighted the potential impact of long-term climate variability on livelihoods and food security. More recent events attest to the continued sensitivity. For example, during the 2000 drought, peanut revenues declined from 68.4 to 17.4 billion FCFA representing a 74% decline. In the same year, revenue from millet and sorghum fell from 30 to 12 billion CFAF, representing a 60% decline in revenue (Mbow et al., 2008; World Bank, 2011b).

The immediate effects of the drought are the migration of rural populations to the cities. Consequently, the growth of urban populations has been relatively rapid. The rural share of the national population fell from 77.1% in 1960 to 70% in 1970 and 60% in 1988 according to the national census.

1.3.2 Floods: The Return of the Rains and Their Flooding Consequences

The pluviometric decrease brings with it a failure of traditional systems of agricultural production. The precariousness of the urban situation does not automatically imply a return to the village unless the opportunity to earn more money becomes possible. The ties to the village can wear thin but rarely break completely as migrants return from seasonal returns, return travels during religious events, for marriage and baptisms. In the city, insertion is possible with access to professional networks.

Floods account for the unpredictability of climate and urbanization without enforceable planning, particularly in the suburbs of Dakar where 49% of the damage is related to housing. In 2009, 33,000 families in the suburbs of Dakar were flood victims, 11% of the population of the region and 25% of that of the suburbs (Gueye et al., 2015). The density and lack of planning are the main causes of Dakar vulnerability to climate change. The main challenge currently facing Dakar and other Senegalese cities remains the increasing severity of floods and their consequences. This phenomenon is partly due to the predominance of informal settlements and thus illegal occupation unfit for habitation and deficiencies in the design and implementation of urban planning documents flood sites.

Floods are one of the most serious hazards in Senegal and a major concern of the government for the past three decades. They affect both urban and rural areas. In 30 years (1980–2009 inclusive), they have affected more than 900,000 people, killed 45 people and caused damage estimated at more than 142 million USD. In peri-urban areas of Dakar, floods have become a regular reality which directly affects the poor. Populations settled in floodplains during periods of drought and lower rainfall than we see recently in a wetter cycle. The total cost of the 2009 floods is estimated at 104 million USD, of which almost 56 million USD comprised direct damage and 48 million USD indirect financial losses (Gueye et al., 2015).

Meanwhile, in recent years floods appear to have become more frequent, in association with the increased frequency of heavy rainfall events. Unrestricted expansion of the built environment likely also played a role. During the decades of persistent drought such expansion took place, for example, in low-lying areas of the Greater Dakar region where runoff has a tendency to collect after heavy rains (Mbow et al., 2008). Between 2000 and 2012, flood damage was recorded in at least 8 years. In three of these years, floods have affected over 150,000 people (179,000 people in 2002, 250,000 people in 2008, and 360,000 people in 2009) and have destroyed infrastructure, public equipment, and private property. In addition, water-borne diseases are said to have become more common due to the presence of stagnant water (World Bank, 2011b).

More recently, recurrent floods across the country have also affected livelihoods. Between 1980 and 2012, floods have affected between 400,000 and 600,000 people annually and have caused a cumulative damage of over 42 million USD. Floods affect the poorest populations, who tend to live in areas with low access to infrastructure. The 2008 floods, for example, affected over 250,000 people and destroyed infrastructure, public equipment and private property. In addition, water-borne diseases also became more common due to the presence of stagnant water (World Bank, 2011b). The recent 2012 floods along the Senegal River and in low-lying areas of Greater Dakar have affected almost 265,000 people, exacerbating the drought-induced food security crisis of 2011–2012 (World Food Programme, 2012). Consecutive floods and droughts, often occurring in the same agricultural season, contribute to the degradation of livelihoods in the longer-run, in the absence of adaptation mechanisms. Climate-related shocks such as droughts and floods affect the ability of households to meet their food requirements, and can lead to food insecurity. Disaster reports released by the Office for Coordination of Humanitarian Affairs (OCHA) ReliefWeb reveal that, over the last 25 years, floods occurring during the rainy season (normally between June and August) have become more frequent.

Historically, regional climate change was linked to persistent drought in the 1970s and the 1980s. More recently, observations suggest a reversal of these conditions with higher precipitation; however, this increase has resulted from increasing rainfall intensity rather than frequency. These trends highlight that rainfall patterns are highly erratic in Senegal, and in the Sahel, more generally, and can result in a more uncertain risk environment in the future. Acknowledging these relationships, this analysis highlights potential linkages between climate and food security, especially children's malnutrition at the national level.

1.4 Implications of Climate Trends in Food Security and Children's Malnutrition

The above discussion depicts a combination of recent rainfall variations and air temperature increases. While the rainfall increases from the 1980s to 1990s helped improve yields, these yield increases have been offset by large declines in the amount of farmland per person. For the past 10 years, however, rainfall has remained steady, and it remains to be seen if the earlier rainfall increases will persist. The sea surface temperature gradient of the Atlantic Ocean swings slowly from north to south on a time scale of decades, and a reversal of the current state could lead to another precipitous Sahelian rainfall decline, a decline augmented by the effects of warmer air temperatures (Funk et al., 2012). Given the potential for such a decline, raising yields in wetter areas may be a more viable option than extending agriculture into more marginal areas. The rapid population expansion in Thies, Diourbel, and Kaolack, heavily populated regions that are important crop producing areas, combined with further drying and warming, could lead to increased food insecurity in these regions (Funk et al., 2012).

In 2023, the estimated population of Senegal was 18.4 million (CIA, 2023). Senegal has a population growth rate of 2.52%; at this rate the population will double every 27 years. Between 1990 and 2010, the population of Senegal increased 64%, with the largest increases in population occurring in Dakar (+1 million), Thies (+0.6 million), and Kaolack (+0.5 million). This population expansion will place increasing stress on limited natural resources. Analysis of crop statistics from the Food and Agriculture Organization of the United Nations (CIA, 2023; FAO, 2023) suggests that increases in crop yields have not kept pace with population growth. Between the 1960s and 2000s, the amount of farmland per person has declined by 300% (from ~0.3 ha per person to ~0.1 ha per person), while yields have only increased by nearly 70%. A continuation of these trends will be unfavorable for national food security; projections for 2025 based on these trends suggest that Senegal will produce 30% less cereal crops per person (Funk et al., 2012). Access to markets is critical to food security in Senegal. In this context, two inter-related climatic impacts could affect the ability of households to access food during critical months. First, if the lean season intensifies due to climatic variability (for example, the onset of the rainy season could be delayed) it is likely that households would have to purchase more of their food during the lean season. Second, under a scenario of increasingly erratic rainfall, extreme climate events such as floods could destroy or severely affect infrastructure, reducing the ability of households to purchase food. Infrastructure is a key aspect, especially in the southern parts of the country where access to markets is critical for ensuring household food security (Maguèye & Ndao, 2013). Malnutrition in childhood has many adverse consequences for child survival and long-term well-being. Climate variability and change stand to exacerbate malnutrition through effects on both the natural and social systems upon which humans depend, including food systems. In areas most highly affected by such changes but with the least adaptive capacity, such as West Africa, the deleterious effects of these changes are most likely to be felt, especially by the most vulnerable (Barros et al., 2014; Ouedraogo et al., 2018; Serdeczny et al., 2017). Ultimately, climatic changes and variations negatively impacting iron-rich meat sources such as livestock, poultry, and fish production, as previously described in this review, therefore have serious health and nutritional implications for children due to decreased access to this essential micronutrient.

In addition to this, water shortages such as during climatic shocks like droughts place severe limitations on children's health, and precipitation and temperature variation are paramount in determining diarrheal prevalence in children under three in Africa generally (Bandyopadhyay et al., 2012), as well as in Senegal.

References

Bandyopadhyay, S., Kanji, S., & Wang, L. (2012). The impact of rainfall and temperature variation on diarrheal prevalence in Sub-Saharan Africa. *Applied Geography,33*, 63–72. https://doi.org/10.1016/j.apgeog.2011.07.017 [23]

Barros, V. R., Field, C. B., Dokken, D. J., Mastrandrea, M. D., & Mach, K. J. (Eds.). (2014). Africa. In *Climate Change 2014: Impacts, adaptation and vulnerability* (pp. 1199–1266). Cambridge University Press. https://doi.org/10.1017/CBO9781107415386.002 [20]

Central Intelligence Agency. (2023). Senegal. In The world factbook. Retrieved from https://www.cia.gov/the-world-factbook/countries/senegal/ [17]

FAO. (2015). Senegal: Socio-economic context and role of Agriculture. [5]

FAO (Food and Agriculture Organization of the United Nations). (2023). FAOSTAT. Retrieved December 15, 2023, from http://faostat.fao.org [18]

Fall, A. S. (1998). Migrants' long-distance relationship and social network in Dakar. *Environment and Urbanization, 10*(April (1)), 135–145. [11]

Funk, C. C., Rowland, J., Adoum, A., Eilerts, G., Verdin, J., & White, L. (2012). A climate trend analysis of Senegal (Fact Sheet 2012–3123). https://doi.org/10.3133/fs20123123 [16]

Gueye, C., Fall, A., & Tall, S. (2015). Dakar, Touba and the Senegalese cities network produced by climate change. *Current Opinion in Environmental Sustainability,13*. https://doi.org/10.1016/j.cosust.2015.02.009 [12]

Hoerling, M., Hurrell, J., Eischeid, J., & Phillips, A. (2006). Detection and attribution of twentieth-century northern and Southern African rainfall change. *Journal of Climatology, 19*, 3,989–4,008. [9]

Huq, S., Reid, H., Murray, L., Kovats, S., Mayers, J., Winkler, H., & Janssen, R. (2006). Climate change and development consultation on key researchable issues: Sectoral issues. [10]

International Institute of Tropical Agriculture. Agroecological Zones. Retrieved July 07, 2020, from http://csi.maps.arcgis.com/apps/MapSeries/index.html?appid=7539d22ab46147ce9888589aea4b1a11 [6]

Lange, S. (2016). EartH2Observe, WFDEI and ERA-Interim data merged and bias-corrected for ISIMIP (EWEMBI). GFZ Data Service, Potsdam. https://doi.org/10.5880/pik.2016.004 [7]

Maguèye, M., & Ndao, M. (2013). Climate risk and food security in Senegal: Analysis of climate impacts on food security and livelihoods. [19]

Mbow, C., Diop, A., Diaw, A. T., & Niang, C. I. (2008). Urban sprawl development and flooding at Yeumbeul suburb (Dakar-Senegal). *African Journal of Environmental Science and Technology, 2*, 75–88. [13]

Mcsweeney, C., New, M., Lizcano, G., & Lu, X. (2010). The UNDP climate change country profiles: Improving the accessibility of observed and projected climate information for studies of climate change in developing countries. *Bulletin of the American Meteorological Society,91*, 157–166. https://doi.org/10.1175/2009BAMS2826.1 [8]

Observatory of Economic Complexity. (2021). Senegal. Retrieved December 01, 2021, from https://oec.world/en/profile/country/sen [4]

Ouedraogo, I., Diouf, N. S., Ouedraogo, M., Ndiaye, O., & Zougmoré, R. B. (2018). Closing the gap between climate information producers and users: Assessment of needs and uptake in Senegal. *Climate, 6*(13). https://www.mdpi.com/2225-1154/6/1/13 [21]

Serdeczny, O., Adams, S., Baarsch, F., Coumou, D., Robinson, A., Hare, W., Schaeffer, M., Perrette, M., & Reinhardt, J. (2017). Climate change impacts in Sub-Saharan Africa: From physical changes to their social repercussions. *Regional Environmental Change, 17*(6), 1585–1600. https://doi.org/10.1007/s10113-015-0910-2 [22]

USAID. (2020). Nutrition Profile: Senegal. Retrieved from https://2017-2020.usaid.gov/sites/default/files/documents/1864/Senegal-Nutrition-Profile-Mar2018-508.pdf [124]

World Bank. (2011b). Climate risk and adaptation country profile: Senegal. World Bank. [14]

World Bank. (2020). World Development Indicators. Retrieved December 03, 2022, from https://databank.worldbank.org/source/world-development-indicators. [3]; Worou, O. N., Tondoh, J. E., Sanou, J., Gaiser, T., Nikiema, P. M., Bayala, J., Bazié, P., Ky- Dembele, C., & Kalinganiré, A. (2018). Intensifying maize production under climate change scenarios in central West Burkina Faso. In W. L. Filho (Ed.), *Handbook of climate change resilience* (pp. 1–23). Springer International Publishing. [51]

World Bank. (2022a). World Bank Open Data. Retrieved December 01, 2022, from https://data.worldbank.org [1]

World Bank. (2022b). World Development Indicators. Retrieved 01, 2022, from https://databank.worldbank.org/source/world-development-indicators [2]

World Food Programme (2012) Food Security Monitoring System (SSSAS—Système de suivi de la sécurité alimentaire au Sénégal). WFP. [15]

Chapter 2
Predicted Cereal Production in Senegal for 2050 and 2080

2.1 Introduction

Senegal is highly vulnerable to shocks associated with climate change. Crop production in Senegal, as mentioned in the previous chapter, is subject to historical droughts events and recent return of the rains and their flooding consequences. These climate-related events have led to the increased vulnerability to food security in Senegal. For instance, the 2011–2012 drought-induced food insecurity and subsequent floods of 2012 in the Sahel was a clear demonstration of climate impacts on food insecurity and children's malnutrition. Despite some development progress in recent years, Senegal is a highly food insecure country: estimates suggest that over 15% of rural households, and over 8% of urban households are food insecure (World Food Programme, 2012). High rainfall variability at inter-annual and inter-decadal scales, particularly in key production areas, affects agricultural production (Fall et al., 2006; McSweeney et al., 2010). The drought of the 1970s and 1980s highlighted the potential of climate-related processes to affect food security in Senegal, and the Sahel more generally (Spinage, 2012).

Groundnut, rice, millet, sorghum, and maize are the main crops grown in Senegal. Crop production and yields for those common crops in Senegal have almost doubled between 1961 and 2010, but with large differences between years (FAO, 2012). Area under production has been rather stable in this period with some significant declines in the late 1990s and mid-2000s. Therefore, increases in production appear to be related to increases in yields, rather than due to area expansion—this is particularly true for paddy rice and maize (FAO, 2012). However, the report mentioned that the total production of the five crops has increased in a non-linear form. Marked increases occurred after the droughts of the early 1970s, and again early 1980s, and more recently after the droughts of 2002 and 2007, suggesting that while crop production is highly sensitive to climate in Senegal, government policies may have adjusted in reaction to drought. Production of groundnut, for instance, decreased significantly in drought years (most notably in 1972, 1977, 1984, 2002, and 2007) (Magueye &

M. Toure, *Agricultural Adaptation to Climate Change in Sub-Saharan Africa*, Advances in African Economic, Social and Political Development,
https://doi.org/10.1007/978-3-031-80716-9_2

Nado, 2013). Rice production, on the other hand, has increased exponentially since the late 1990s due to an increase in both area harvested (especially along the Senegal River Valley) and increases in productivity (Magueye & Nado, 2013).

Since the historical relationship between crop production and rainfall statistics illustrates the importance of climatic variables on food security, one might ask how this relationship will evolve in the future: horizons 2050 and 2080 for instance. In response to this question, the objective of this chapter is to estimate the predicted cereal production in Senegal in 2050 and 2080 based on the scenario of changes not only on climate variables but also changes in area and yields of rice, maize, sorghum, and pearl millet. The contribution of this chapter is to quantify the impact of climate change on major cereal crops production—rice, maize, sorghum, and pearl millet—in the near and long term. For that, we used the same study sites and data sources from Jha et al. (2021), which quantified the impacts of COVID-19 on major cereal crops production and their cascading impact on Senegalese economy. We expanded their work to predict the production of major cereal in Senegal in 2050 and 2080.

This chapter is organized into six sections. In Sect. 2.2, data collection and materials used for this study are explained in great detail. This includes the study sites and data sources for weather, soil, area, and yield data. After that, Sect. 2.3 discusses the methodology, which presents the crop model as well as the model calibration and evaluation. Scenarios of change in area and yield due to climate change are presented in Sect. 2.4. Assumptions and limitations of some of the data sources can be found in Sect. 2.5, while Sect. 2.6 focuses on the evaluation and the statistical analysis of this study. This chapter ends with the presentation of the simulated cereal production in 2050 and 2080 after exposing the model calibration and validation as well as the predicted crop yields.

2.2 Data and Materials

2.2.1 Study Sites

The study was conducted at three different locations which represent different agro ecological zones: northern arid and semi-arid, central sub-humid, and southern humid in Senegal, which are based on low to high rainfall distributions respectively (Table 2.1 and Fig. 2.1). The major farming system in the northern arid region is agro pastoral livestock production with groundnut (Arachis hypogaea L.) and pearl millet (Pennisetum glaucum L.) as the major crops; in the central sub-humid region cotton (Gossypium hirsutum L.), pearl millet (Pennisetum glaucum (L.) R. Br.), sorghum (Sorghum bicolor L. Moench), and maize (Zea mays L.); and in the southern humid region rice (Oryza sativa L.), pearl millet and maize. Across the country the major cereals crops grown are pearl millet, rice, maize, and sorghum (Leippert et al., 2020). Therefore, these four major cereal crops were chosen for the impact assessment.

Table 2.1 Average annual precipitation (mm) of the study sites for 1990–2020 in Senegal

Geographical location	Name of study site	Average annual precipitation (mm) from 1990 to 2020
North Arid	Yang Yang	457.11
Central Sub-Humid	Tambacounda	663.71
South Humid	Guiragor	2834.93

Source Author's computations based on NASA Langley Research Center data

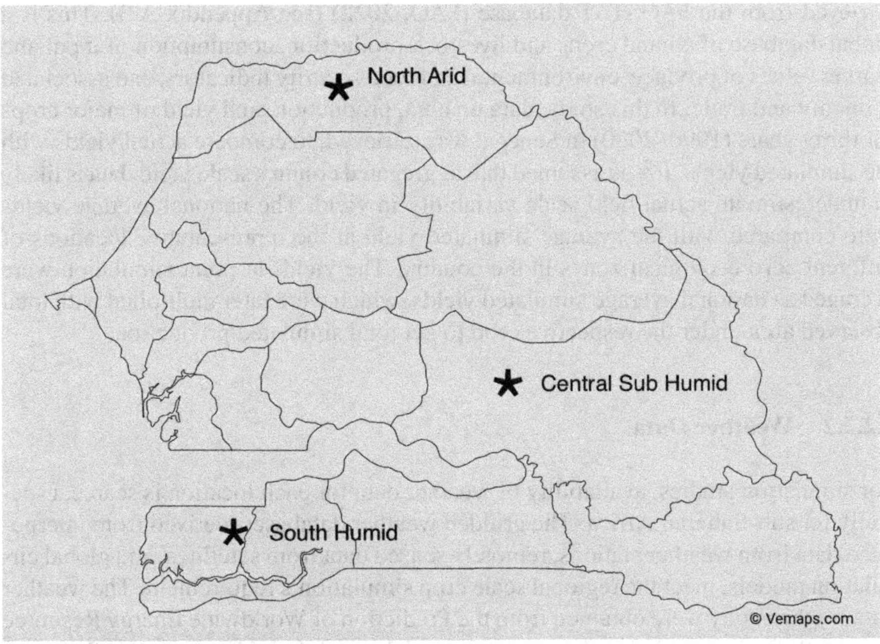

Fig. 2.1 Senegal's study sites. *Source* "Projecting potential impact of COVID-19 on major cereal crops in Senegal and Burkina Faso using crop simulation models" by Jha et al. (2021)

The locations were Yang Yang (northern arid; 15.66°N latitude, 15.42°W longitude, and elevation of 40 m above mean sea level), Tambacounda (central sub-humid; 13.77°N latitude, 13.67°W longitude, and elevation of 24 m above mean sea level) and Guiragor (southern humid; 13.01°N latitude, 16.07°W longitude, and elevation of 17 m above mean sea level).

2.2.2 *Data Sources*

2.2.2.1 Actual Areas and Yields

Large-scale studies on crop yield variability require long-term regional or national data to validate any simulation studies (Grassini et al., 2015; Van Wart et al., 2015). Aggregated country level data on actual harvested area, production, and yield for major cereal crops (for instance: rice, maize, sorghum, and pearl millet) were retrieved from the FAOSTAT database (FAO, 2022) (see Appendix A.3). This is a global database of annual crops and livestock production, consumption of input and market values of produce, environmental and food security indicators, and associated economy and trade. In this study, data on area, production, and yield of major crops for thirty years (1990–2020) in Senegal was retrieved, to compare actual yields with the simulated yields. It was assumed that aggregated country scale yield data is likely to underestimate actual field scale variability in yield. The national average yields were compared with the average simulated yield at the representative locations of different agro ecological zones in the country. The yields at point simulation were averaged as national average simulated yields, which were later multiplied with total observed area under the respective crop to get total simulated production.

2.2.2.2 Weather Data

For simulation studies, availability of weather data for each location is scarce, especially for sub-Saharan Africa. The gridded weather databases, derived from interpolated data from weather stations, remotely sensed data from satellites, and global circulation models, meet the regional scale crop simulation's requirement. The weather data for this study were obtained from the Prediction of Worldwide Energy Resource (POWER) Project of National Aeronautics and Space Administration's (NASA), produced by the NASA Langley Research Center. NASA-POWER provides data on a global grid with a spatial resolution of 0.5° latitude by 0.5° longitude (https://power.larc.nasa.gov/data-access-viewer/). The gridded daily weather data from NASA-POWER are derived from remotely sensed satellite observations coupled with the land surface model, which is supported by data from ground stations and satellites.

These reanalyzed gridded weather datasets have been used for yield simulations at regional scale (Aboelkhair et al., 2019; Bai et al., 2010; Lobell et al., 2010; Parkes et al., 2019). Thirty-year (1990–2020) weather parameters such as daily solar radiation, maximum and minimum temperature, and rainfall at each of the locations in Senegal were downloaded from NASA-POWER. The long-term rainfall trend remains the same which supports vegetation distribution till date (Ilboudo et al., 2020). The weather data were ingested to the crop model software used for this analysis, which is the Decision Support System for Agrotechnology Transfer (DSSAT v4.8; Hoogenboom et al., 2017). This crop model, which we will talk more about in detail in the next section, uses WeatherMan to convert raw data to DSSAT weather format (Pickering et al., 1994).

2.2.2.3 Soil Data

In this study, we used 10 km resolution gridded global soil profile datasets (Han et al., 2019), which are synthesized using SoilGrids1km datasets (Hengl et al., 2014) and HarvestChoice soil profile: HC27; (Koo & Dimes, 2010), which are compatible with the DSSAT format. Based on the coordinates of the locations, the soil input was ingested to the soil database of DSSAT. The soil fertility factor (SLPF) value was adjusted and modified manually for different crops by minimizing the error between the observed and simulated yield. We used 0.9 for rice (de Vries et al., 2012), 0.5 for maize (Amouzou et al., 2018), 0.6 for sorghum (Singh et al., 2014), and 0.5 for pearl millet (Singh et al., 2017) in this study. The detailed soil characteristics are shown in Appendix A.1 (Fig. A.1).

2.3 Methodology

2.3.1 Model Calibration and Evaluation

Crop simulation models have been used to study growth and development of crops under different environments (Araya et al., 2017; Jha et al., 2018, 2021). The crop model used for this study was DSSAT version 4.8, which simulates crop growth and development based on the interaction among genotype, environment, and management practices (Elias et al., 2016; Hoogenboom et al., 2017) using daily weather data and soil inputs. Crop genetic coefficients are the parameters which determine cultivar growth and development under different environments. The cultivars which were already calibrated and evaluated for Western African regions were selected for this study. The cultivar coefficients for rice (Modified IR 64) (de Vries et al., 2012), maize (Massongo) (Worou et al., 2018), sorghum (CSM-335) (Singh et al., 2014), and pearl millet (CIVT) (Singh et al., 2017) were used for simulation (see Appendix A.2, Tables A.1, A.2, A.3 and A.4). The observed yield for 2020 (FAO, 2022) was considered as the target to evaluate the model. The models were calibrated by adjusting the photosynthesis factor (SLPF) in the soil input file to improve model performance. For calibration, best available conditions (Southern Humid locations) were preferred (Lobell et al., 2009). The model was first calibrated for 2020 using weather data for 2020 and observed yield from FAO as the target yield and then evaluated for long-term observed yield data from FAO.

The calibrated model was used to simulate long-term crop yield (1990–2020) for all three locations in Senegal. Soil characteristics, crop genotype and management practices at the individual locations were used. Crop yields were simulated using different daily weather data corresponding to the specific year. The national simulated average annual yield (1990–2020) under two different scenarios; with and without fertilizer application, were compared with the national observed yield from FAO (2022) (see Figs. 2.4 and 2.5). After model evaluation from 1990 to 2020, the yields were simulated for 2050 and 2080, using projected climate data to be discussed in Sect. 2.4, at different locations and converted to the national average yields. The country level simulated production was estimated through a simple approach of multiplying different scenarios of national average simulated yields with the different scenarios of change in cultivated area from a reference value. The reference value of cultivated area of different crops was retrieved from the latest total cultivated area of the crops in 2020 (FAO, 2022) and the reference value for yield was a national average 2020 simulated yield.

2.3.2 *Crop and Management Information*

Most of Senegal lies in the arid and semi-arid tropics with variable spatiotemporal rainfall distribution and low soil fertility, which leads to poor agricultural productivity in these low input rainfed systems. Management practices such as planting dates, plant population density, fertilizer timing, and application for each crop were retrieved from reported literature in Senegal (Khairwal et al., 2007; Ekeleme et al., 2008; Sommer et al., 2013; Okuyama et al., 2017). To simulate yields of the four major cereal crops (rice, maize, sorghum, and pearl millet) under rainfed conditions, the planting date was considered as June 15th, which is an average date for the onset of the rainy season in Senegal (Sane et al., 2018). For uniformity, the fertilizer application rate (100 kg/ha urea–nitrogen) with split dose of nitrogen at sowing and 40 Days After Sowing (DAS) was kept constant for all crops. The row spacing was kept constant for maize at 3 cm; sorghum at 1 cm; pearl millet at 3 cm and for rice at 1 cm, for all locations. The population density was kept constant at 80 seeds/m^2, 10 seeds/m^2, 5 seeds/m^2, and 150 seeds/m^2 respectively for maize, sorghum, pearl millet, and rice for all locations.

2.3.3 *Crop Model: Decision Support System for Agrotechnology Transfer (DSSAT)*

2.3.3.1 Introduction

Crop models offer the benefit of increasing our understanding of crop responses to management in different soil and climatic conditions. They can also provide insights in what might happen to productivity under various climate change scenarios, a domain beyond the reach of field experimentation. The outputs can inform key decision makers at local, national, and regional levels in order to put the appropriate measures in place. These models calculate expected growth and development based on equations that describe how a crop, as a community of plants, responds to soil and weather conditions. At their simplest level of interpretation, the equations used in a model are a set of differential equations representing rates of growth or development. Numerical integration over time, typically with daily or hourly time steps, allows estimation of growth, development, and water and nutrient levels. The equations are based on information from crop physiology, soil science, meteorology, and other fields.

The Decision Support System for Agrotechnology Transfer (DSSAT) is a software application program that comprises crop simulation models for over 42 crops (as of version 4.8) as well as tools to facilitate effective use of the models. It has a collection of independent programs that operate together; crop simulation models are at its center. Databases describe weather, soil, experiment conditions and measurements, and genotype information for applying the models to different situations. Software helps users prepare these databases and compare simulated results with observations to give them confidence in the models or to determine if modifications are needed to improve accuracy. We used the DSSAT to simulate monocrop production systems considering weather, genetics, soil water, soil carbon and nitrogen, and management.

2.3.3.2 Overall Description

The DSSAT-CSM (Cropping System Model) simulates growth, development, and yield of a crop growing on a uniform area of land under prescribed or simulated management as well as the changes in soil water, carbon, and nitrogen that take place under the cropping system over time. The DSSAT-CSM has a main driver program, a land unit module, and modules for the primary components that make up a land unit in a cropping system (Fig. 2.2). The Primary modules are for weather, soil, plant, and management components. Collectively, these components describe the time changes in the soil and plants that occur on a single land unit in response to weather and management.

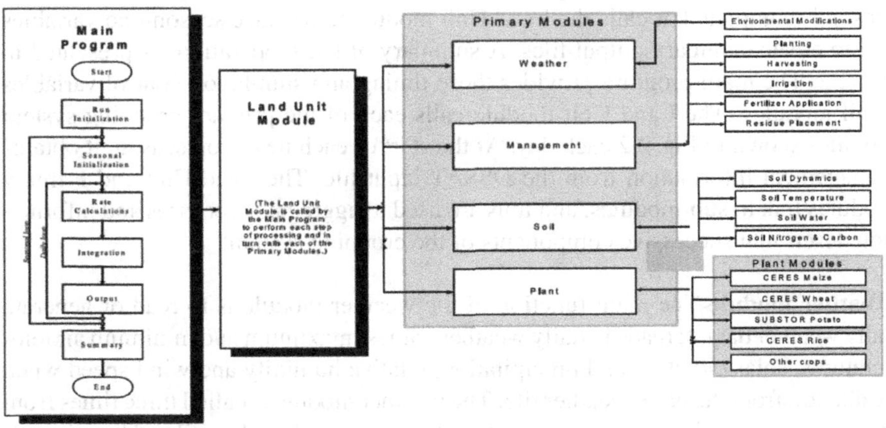

Fig. 2.2 Overview of the components and modular structure of the DSSAT. *Source* "The DSSAT cropping system model" by Jones et al. (2003)

Each module has six operational steps, as shown in Fig. 2.2 (run initialization, season initialization, rate calculations, integration, daily output, and summary output). The main program controls when each of these steps is active, and when each module performs the task that is called for. This feature allows each module to read its own inputs, initialize itself, compute rates, integrate its own state variables, and write outputs completely independent from the operation of other modules. State variables are written after integration to represent the state of the system at the end of the day, and initial values are written during initialization for day 0. Different types of applications are accomplished in DSSAT-CSM by using different modes to call the land unit module on a daily basis; the mode is specified as a command line argument when the model is run. The basic mode provides for interactive sensitivity analysis and comparison of simulated versus observed field data.

Table 2.2 lists the primary and sub modules currently used in the CSM and summarizes their functions. There are two important points to be made about this table. First, sub-modules operate exactly like primary modules. Each sub-module will usually perform six steps. Second, we used the following plant modules: CERES maize, CERES rice, CERES sorghum, and CERES millet as shown in Fig. 2.2.

2.3.3.3 Component Descriptions

The main program reads information from the DSSAT standard file that describes a particular situation to be simulated and sets a number of variables for controlling a simulation run. It initiates the simulation by setting the DYNAMIC variable for initializing the run and calls the Land Unit module. It then starts a crop season time loop and calls the Land Unit module for initializing variables that must be set at the start of each season. After initialization of the seasonal loop, the main program starts a daily loop and calls the Land Unit module three times in sequence, first to compute rates, secondly to integrate, and finally to report daily outputs. After a crop season is completed, it calls the Land Unit module to produce season-end variables and to create summary output files. A summary of these operations is presented in Fig. 2.2. The main program provides these timing and simulation control variables to all modules. The Land Unit module calls each of the primary cropping system modules shown in Fig. 2.2 each day. At the start of each new crop season, it obtains management information from the DSSAT input file. The Land Unit and Primary modules link to sub-modules, and thus are used to aggregate processes and information describing successive components of the cropping system.

Weather module The main function of the weather module is to read or generate daily weather data. It reads in daily weather values (maximum and minimum air temperatures, solar radiation and precipitation, relative humidity and wind speed when available), from the daily weather file. The weather module is called three times from the main program for initialization, rate calculations and to close files. Upon initialization, the weather file is opened. In the rate calculations section, weather values

Table 2.2 Summary description of modules in the DSSAT-CSM

Modules	Sub-modules	Behavior
Main program (DSSAT-CSM)		Controls time loops, determines which modules to call based on user input switches, controls print timing for all modules
Land unit		Provides a single interface between cropping system behavior and applications that control the use of the cropping system. It serves as a collection point for all components that interact on a homogenous area of land
Weather		Reads or generates daily weather parameters used by the model. Adjusts daily values if required, and computes hourly values
Soil	Soil dynamics	Computes soil structure characteristics by layer. This module currently reads values from a file, but future versions can modify soil properties in response to tillage, etc.
	Soil temperature modules	Computes soil temperature by layer
	Soil water module	Computes soil water processes including snow accumulation and melt, runoff, infiltration, saturated flow, and water table depth. Volumetric soil water content is updated daily for all soil layers. Tipping bucket approach is used
	Soil nitrogen and carbon module	Computes soil nitrogen and carbon processes, including organic and inorganic fertilizer and residue placement, decomposition rates, nutrient fluxes between various pools and soil layers. Soil nitrate and ammonium concentrations are updated on a daily basis for each layer
Individual plant growth modules	CERES-Maize; CERES-Wheat; CERESRice; SubStor-Potato; Other plant models	Modules that simulate growth and yield for individual species. Each is a separate module that simulates phenology, daily growth and partitioning, plant nitrogen and carbon demands, senescence of plant material, etc.
Management operations module	Planting	Determines planting date based on read-in value or simulated using an input planting window and soil, weather conditions
	Harvesting	Determines harvest date, based on maturity, read-in value or on a harvesting window along with soil, weather conditions
	Irrigation	Determines daily irrigation, based on read-in values or automatic applications based on soil water depletion
	Fertilizer	Determines fertilizer additions, based on read-in values or automatic conditions
	Residue	Application of residues and other organic material (plant, animal) as read-in values or simulated in crop rotations

Source "The DSSAT cropping system model" by Jones et al. (2003)

Table 2.3 Weather module input variables

Variable name	Definition	Units
DATE	Julian date in YYDDD format	–
PAR	Daily photosynthetically active radiation	mol[photon]/m^2 day
RAIN	Daily rainfall	mm
SRAD	Daily solar radiation	MJ/m^2
TMAX	Daily maximum temperature	°C
TMIN	Daily minimum temperature	°C

Source "The DSSAT cropping system model" by Jones et al. (2003)

are read into the model on a daily basis from file WEATHER.INP. Table 2.3 lists the weather input data read. The close section is invoked to close the weather input file. For our study, we uploaded daily weather data from 1990 to 2020 reported by NASA-Power website reported in our three study sites (Yang Yang, Tambacounda, and Guiragor) and we also uploaded predicted daily weather data for 2050 and 2080 in those same study sites reported by MarkSim DSSAT weather file generator.

Soil module The soil in the land unit is represented as a one-dimensional profile; it is homogenous horizontally and consists of a number of vertical soil layers. The soil module integrates information from four sub modules: soil water, soil temperature, soil carbon and nitrogen, and soil dynamics. Table 2.4 shows the list of variables found in the soil module. In our study, we used 10 km resolution gridded global soil profile datasets (Han et al., 2019), which are synthesized using SoilGrids1km datasets (Hengl et al., 2014) and HarvestChoice soil profile: HC27; (Koo & Dimes, 2010), which are compatible with the DSSAT format. Based on the coordinates of the locations, the soil input was ingested to the soil database of DSSAT. The soil input values for our three study sites can be found in Appendix A, Fig. A.1 Table 2.5 defines the variables produced by this module for use in other modules. The soil dynamics module is designed to read in soil parameters for the land unit and to modify them based on tillage, long-term changes in soil carbon, or other field operations. The soil dynamics module currently reads in soil properties from a file (as Appendix A, Fig. A.1).

Plant module The plant growth module computes crop growth and development based on daily values of maximum and minimum temperatures, radiation and the daily value of soil output values. This module also simulates leaf area index (LAI), which is used in the soil water module to compute evapotranspiration. Table 2.6 shows the list of input variables of the plant module. To simulate the plant growth of our four crops: rice, maize, sorghum, and pearl millet, we used the CERES models for which the plant life cycle is divided into several phases. Rate of development is governed by thermal time, or growing degree days (GDD), which is computed based on the daily maximum and minimum temperatures. For any given day d of the growing season (defined here as the period between the planting day t_p, and the crop maturity

Table 2.4 Soil module input variables

Variable name	Definition	Unit
SCOM	Soil color (Munsell color system)	–
SALB	Albedo	–
SLU1	Evaporation limit	mm
SLDR	Drainage rate	fraction day^{-1}
SLRO	Runoff curve number	–
SLNF	Mineralization factor	0–1 scale
SLPF	Photosynthesis factor	0–1 scale
SMHB	pH in buffer determination method	–
SMPX	Extractable phosphorus determination code	–
SMKE	Potassium determination method	–
SLMH	Master horizon	–
SLLL	Lower limit of plant extractable soil, water, or wilting point	$cm^3\ cm^{-3}$
SDUL	Drained upper limit, or field capacity	$cm^3\ cm^{-3}$
SSAT	Saturated upper limit	$cm^3\ cm^{-3}$
SRGF	Root growth factor	0–1 scale
SSKS	Saturated hydraulic conductivity	cm h^{-1}
SBDM	Bulk density (moist)	g cm^{-3}
SLOC	Soil organic carbon concentration	%
SLCL	Clay (<0.002 mm)	%
SLSI	Silt (0.05–0.002mm)	%
SLCF	Coarse fraction (>2 mm)	%
SLNI	Total nitrogen concentration	%
SLHW	pH in water	–
SLHB	pH in buffer	
SCEC	Cation exchange capacity	cmol (+) kg^{-1}

Source "The DSSAT cropping system model" by Jones et al. (2003)

t_m), or for any d in $[t_p, t_m]$, the thermal time (TT) is computed from the daily mean temperature (tas, in °C) by using a reference temperature (T_0):

$$\mathrm{TT}(d) = \mathrm{tas}(d) - T_0 \tag{2.1}$$

$$\mathrm{GDD}(d) = \sum_{i \le d} \max(0, TT(i)) \tag{2.2}$$

(Appendix C shows how GDD and daily growth rates are derived). The GDD required to progress from one growth stage to another are defined as a user input. The genetic coefficients for the DSSAT CERES-Maize, CERES-Rice, CERES-Sorghum,

Table 2.5 Soil module output variables

Variable	Definition	Variable	Definition
BD(L)	Bulk density, soil layer L (g[soil]/cm^3[soil])	PH(L)	pH in soil layer L
DLAYR(L)	Thickness of soil layer L (cm)	SAT(L)	Soil water content at saturation (cm^3[water]/cm^3[soil])
DUL(L)	Soil water content at drained upper limit in soil layer L (cm^3[water]/cm^3[soil])	SRFTEMP	Temperature of surface litter (°C)
LL(L)	Soil water content in soil layer L at lower limit of plant extractable soil water (cm^3[water]/cm^3[soil])	ST(L)	Soil temperature in soil layer L (°C)
NH_4(L)	Ammonium N in soil layer L (g[N]/g[soil])	SW(L)	Soil water content in layer L (cm^3[water]/cm^3[soil])
NLAYR	Actual number of soil layers	SWDELTS(L)	Change in soil water content due to drainage in layer L (cm^3[water]/cm^3[soil])
NO_3(L)	Nitrate in soil layer L (g[N]/g[soil])	WINF	Water available for infiltration-rainfall plus net irrigation minus runoff (mm/d)

Source "The DSSAT cropping system model" by Jones et al. (2003)

and CERES-Pearl millet used in our study can be found in Appendix A.2 (Tables A.1, A.2, A.3 and A.4). Cultivar-specific inputs for all DSSAT-CSM CERES models are presented in absolute terms as shown in Appendix A.2 (Tables A.1, A.2, A.3 and A.4). Table 2.7 shows the list of output variables derived by the plant module.

Management module The management module determines when field operations are performed by calling sub-modules. Currently, these operations are planting, harvesting, applying inorganic fertilizer, irrigating and applying crop residue and organic material (as shown in Fig. 2.2). These operations can be specified by users in the standard 'experiment' input file. Users specify whether any or all of the operations are to be automatic or fixed based on input dates or days from planting.

Harvesting can occur on given dates, when the crop is mature, or when soil water conditions in the field are favorable for machine operation. Irrigation can be applied on specific dates with specified irrigation amounts. Similarly, fertilizer can be applied on fixed dates in specified amounts, or the applications can optionally be controlled by plant needs for nitrogen via the nitrogen stress variable from the plant module. Crop residue and organic fertilizer, such as manure, is applied either at the start of simulation, after harvesting the crop or on fixed dates similar to inorganic fertilizer applications. These management options allow users a great deal of flexibility for simulating experiments. Table 2.8 displays the output variables generated by the management module.

Table 2.6 Plant module input variables

Variable name	Definition	Units
EMP1	Empirical coefficient for LAI computation, maximum leaf area expansion per leaf	m^2/leaf
EMP2	Empirical coefficient for LAI computation	–
fc	Fraction of total crop growth partitioned to canopy	–
intot	Duration of reproductive stage	degree-days
lai	Leaf area index	m^2/m^2
Lfmax	Maximum number of leaves	–
n	Leaf number	–
nb	Empirical coefficient for LAI computation	–
p1	Dry matter of leaves removed per plant per unit development after maximum number of leaves is reached	g
PD	Plant density	plants/m^2
rm	Maximum rate of leaf appearance	leaf/day
sla	Specific leaf area	m^2/g
tb	Base temperature above which reproductive growth occurs	°C
w	Total plant dry matter weight	g/m^2
wc	Canopy dry matter weight	g/m^2
wr	Root dry matter weight	g/m^2

Source "The DSSAT cropping system model" by Jones et al. (2003)

Table 2.7 Plant module output variables

Variable	Definition	Variable	Definition
NSTRES	Nitrogen stress factor (1 = no stress, 0 = max stress)	UNO3(L)	Rate of root uptake of NO^3(kg[N]/ha d)
RLV(L)	Root length density for soil layer L (cm[root]/cm^3[soil])	XHLAI	Healthy LAI (m^2[leaf]/m^2[ground])
SENCLN(I,J)	Daily senesced plant matter. I = 0 for surface, 1 for soil; J = 1 for C, 2 for lignin, 3 for N (g[C, N, or lignin]/m^2d))	XLAI	LAI (m^2[leaf]/m^2[ground])
STDGOY(I)	Day when plant stage I occurred (YYDDD)/(YRDOY)	YREMRG	Day of emergence (YYDDD)
UNH4(L)	Rate of root uptake of NH_4(kg[N]/ha d))	YRNR8	Harvest maturity date (YYDDD)/YRDOY)

Source "The DSSAT cropping system model" by Jones et al. (2003)

Table 2.8 Management module output variables

Variable	Definition	Variable	Definition
FERTYPE	Fertilizer type for current application	RESSRF	Residue left on surface of soil (kg[residue]/ha)
FERDEPTH	Fertilizer incorporation depth on current day of simulation (cm)	RESNIT	N concentration of the residue for current application (%)
FERNIT	Amount of nitrogen in fertilizer applied on current day of simulation (kg[N]/ha)	RESDEPTH	Incorporation depth of newly added residues (cm)
HAREND	End of season or harvest date (YYDDD)	RESTYPE	Residue type for current application
IRRAMT	Irrigation amount for today (mm/d)	RESTYPE	Residue type for current application
RESSOL	Amount of residue applied to the soil (kg[dry matter]/ha)	YRPLT	Planting date (YYDDD)

Source "The DSSAT cropping system model" by Jones et al. (2003)

2.3.3.4 Data Requirements

The DSSAT models require the minimum data set for model operation. They encompass data on the site where the model is to be operated, on the daily weather during the growing cycle, on the characteristics of the soil at the start of the growing cycle or crop sequence, and on the management of the crop (e.g., seeding rate, fertilizer applications, irrigations). Required weather data (as shown in Table 2.9) encompass daily records of total solar radiation incident on the top of the crop canopy, maximum and minimum air temperature above the crop, and rainfall. Typical crop management factors include planting date, planting depth, row spacing, plant population, fertilization, irrigation, and inoculation.

2.3.3.5 Limitations of the Model

In order to simulate crop yield via genotype specific parameters (GSPs), the DSSAT uses multi-objective evolutionary algorithms (such as the GLUE tool[1] or the GEN-

[1] The General Likelihood Uncertainty Estimation (GLUE) is a statistical approach that results in multiple sets of parameter values that are equally as likely as the final solution. To estimate the most likely values for the GSPs for a new cultivar, a user first has to provide the required input files associated with weather, soil, and crop management, and basic observations, especially for phenology, yield, and yield components.

Table 2.9 Contents of minimum data sets for operation

Site	Latitude and longitude, elevation; average annual temperature; average annual amplitude in temperature Slope and aspect; major obstruction to the sun (e.g., nearby mountain); drainage (type, spacing and depth); surface stones (coverage and size)
Weather	Daily global solar radiation, maximum and minimum air temperatures, precipitation
Soil	Classification using the local system and (to family level) the USDA-NRCS taxonomic system Basic profile characteristics by soil layer: in situ water release curve characteristics (saturated drained upper limit, lower limit); bulk density, organic carbon; pH; root growth factor; drainage coefficient
Initial conditions	Previous crop, root, and nodule amounts; numbers and effectiveness of rhizobia (nodulating crop) Water, ammonium and nitrate by soil layer
Management	Cultivar name and type Planting date, depth and method; row spacing and direction; plant population Irrigation and water management, dates, methods and amounts or depths Fertilizer (inorganic) and inoculant applications Residue (organic fertilizer) applications (material, depth of incorporation, amount and nutrient concentrations) Tillage Environment (aerial) adjustments Harvest schedule

Source "The DSSAT cropping system model" by Jones et al. (2003)

CALC tool[2]), which are often criticized due to slow convergence and the large number of function evaluations required before an acceptable solution can be found.

Also, another weakness of the DSSAT ecosystem and many other crop modeling systems is the limited capability for handling the impact of biotic stresses caused by insect pests, diseases, and weeds. DSSAT currently has a static system that allows a user to define biotic stressors based on field damage observations. However, there is no coupling with dynamic pest and disease models.

[2] The Genetic Coefficient Calculator or GENCALC uses a rule-based approach to determine the value for one or more GSPs During the calibration process, these GSPs are varied to minimize the error between the simulated and observed trait. GENCALC normally optimizes the phenological GSPs first, followed by growth, and then yield components and yield for final optimization.

2.4 Scenarios of Change in Area and Climate Variables Due to Climate Change

2.4.1 Scenarios of Change in Area

In addition to a widespread unconfirmed speculation about the potential impact of natural disasters—such as droughts, floods, and rainfall variability in Senegal, as mentioned earlier—on input availability due to the disruption of supply chains in Africa in general, and in Senegal in particular, there has also not been any scientific evidence that quantifies how such disruptions in the supply chain may impact crop production and agricultural GDP. Disruptions of input supply chains (e.g., seeds, fertilizers and nutrients, pesticides, availability of credit) can influence farm operations and crop management decisions that can decrease crop yields. In addition, the movement of migrant labor from cities to villages and unavailability of off-farm jobs in rural areas led to reduced earnings in Senegal (Balde et al., 2020). Several may engage in farming, especially those who can or have access to land. Consequently, it could lead to an increase in the overall crop planting area. Due to lack of scientific evidence, it is difficult to quantify how much these disruptions in the supply chain and return of people to villages may impact crop production and overall agricultural GDP.

Based on these assumptions, five different types of hypothetical scenarios of percentage change in the planting area were designed and classified as: –10, –5, 0, +5, +10%. For each hypothetical scenario, we apply the percentage change in 2050 and 2080 with 2020 being the reference year for 2050 and 2050 being the reference year for 2080. For example: for –10% change in planting area, the planting area for 2050 is 10% less than the planting area in 2020 and the planting area for 2080 is 10% less than the planting area in 2050. In the next chapter, we will explore the impact of technical changes, changing quality of soil, and longer persistent disasters on planting areas and crop production.

2.4.2 Scenarios of Change in Climate Variables

In order to predict crop yield in 2050 and 2080, MarkSim DSSAT weather file generator was used. This application uses the well-known MarkSim™ application (Jones & Thornton, 2000; Jones et al., 2002) working off a 30 arc-second climate surface derived from WorldClim (Hijmans et al., 2005). Point and click on the map and up to 99 WTG files are prepared ready for use with DSSAT®. Download and unpack to a directory on your machine and they are ready for use with the DSSAT4 crop modeling system (Blade et al., 2020 Hoogenboom et al., 2019). MarkSim is a weather generator that uses 720 classes of weather, worldwide, to calculate the coefficients of a third-order Markov rainfall generator. This constitutes 'stochastic downscaling' as it fits a Markov model to the GCM output and uses it to generate weather data for the site indicated.

The climate change projections are obtained by forcing the regional climate model by the outputs of the global climate model (GCM) under greenhouse gases emission scenarios RCP 2.6 and RCP 8.5. The RCP 2.6 (or most optimistic scenario) and RCP 8.5 (or most pessimistic scenario) scenarios correspond to emissions of 2.6 W/m^2 and 8.5 W/m^2 of greenhouse gases, respectively.

Given these two climate change scenarios—RCP 2.6 and RCP 8.5—shown in Fig. 2.3, there are going to be two predicted yields for each crop at each study site (rice, maize, millet, and sorghum) in 2050 and in 2080. In order to get the national predicted crop yield for each cereal in 2050 and 2080, we will take the average of the crop yields produced in the three study sites. This means that for each crop, there will be four hypothetical predicted yields (two in 2050 and two in 2080). Ultimately the five hypothetical scenarios of percentage change in planting area (–10, –5, 0, +5, +10%) and the four hypothetical predicted yields for each crop will lead to twenty scenario matrices for each crop.

2.5 Assumptions and Limitations

The simulations were performed under rainfed conditions assuming there was no surface water storage and soil saturation. The soil characteristics were kept static as in SoilGrids1km datasets and there were no biotic and abiotic constraint to the simulated yield, though these conditions are prevalent in African conditions (Stewart et al., 2020; Saito et al., 2019; Ouedraogo & Ayantunde, 2019). Following the model validation for 2020, long-term simulations were performed using the same cultivars irrespective of decadal change of varietal improvement (Walker et al., 2015) and the same fertilizer rates irrespective of change in recommendation (Rware et al., 2014). However, in the next chapter we will explore the impact of the changing quality of soil in crop yields and production.

2.6 Evaluation and Statistical Analysis

We compared predicted and observed yield and used the coefficient of determination (R2; Eq. 2.3), and Root Mean Square Error (RMSE; Eq. 2.4) (Willmott, 1982) to measure the performances of the calibrated model.

$$R^2 = \frac{\left[\sum_{i=0}^{n}(O - \bar{O})(S - \bar{S})\right]^2}{\sum_{i=0}^{n}(O - \bar{O})\sum_{i=0}^{n}(S - \bar{S})^2} \tag{2.3}$$

$$\text{Root Mean Square Error} = \sqrt{\frac{\sum_{1}^{n}(S - O)^2}{n}} \tag{2.4}$$

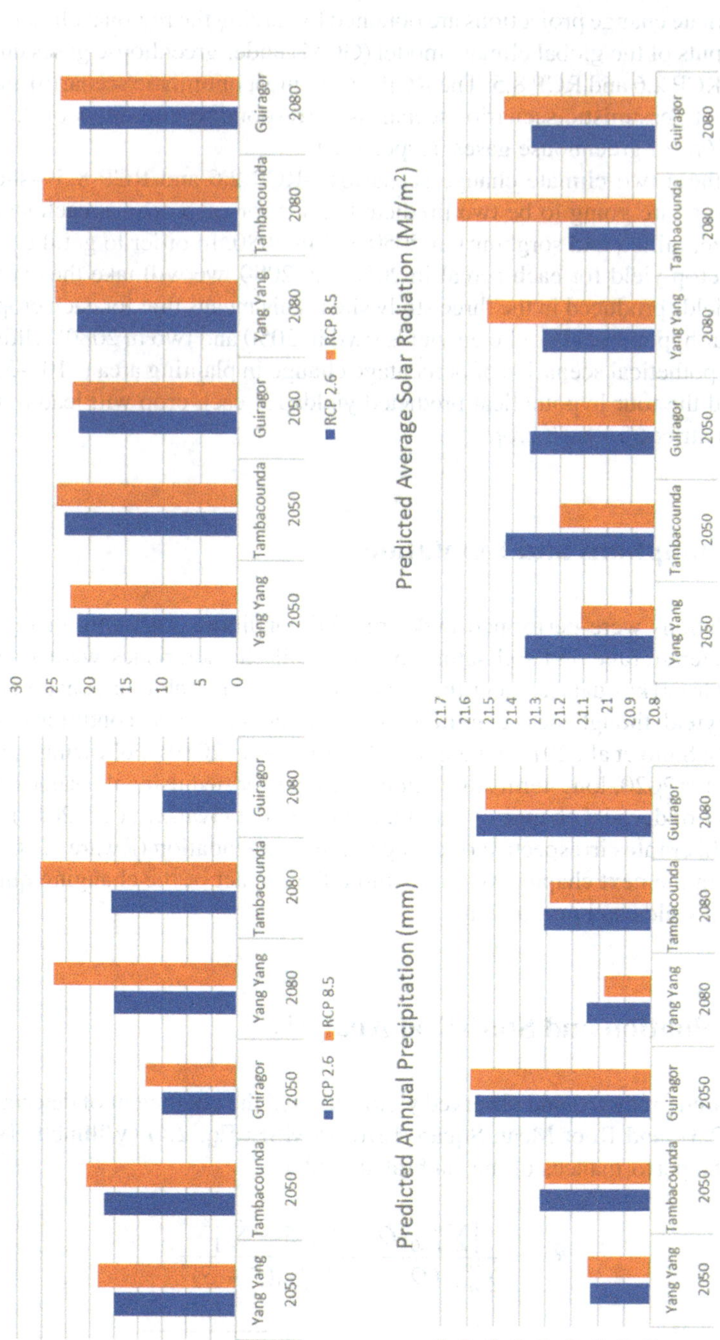

Fig. 2.3 Predicted climate variables in Senegal's three study sites. *Source* Author's computations

where S and O are simulated and observed yields, respectively; $\overline{S}$ and $\overline{O}$ are mean simulated and observed yield respectively; i is variable, and n is number of variables.

2.7 Results and Discussion

2.7.1 Model Calibration and Validation

Initially, the Crop Environment REsource Synthesis (CERES) sub-model in DSSAT was calibrated for all the grain cereals. The model was calibrated for 2020. The calibrated model was used to evaluate for the years 1990–2020 and to simulate yields for 2050 and 2080 at the three study sites. While the yields of rice and sorghum were found to be lowest in the northern arid locations, the yields of maize and pearl millet were found to be lowest in the southern humid regions (as shown in Fig. 2.4). The yield gradient of rice and sorghum seems to follow the annual rainfall availability pattern across the locations in Senegal.

Maize simulated yield hardly captures the long-term observed yield pattern (as shown in Figs. 2.4 and 2.5). Database of the observed yield from FAO (2022) showed that there was 1.6 times increase in cultivated area for maize from 108,114 ha in 2002 to 175,575 ha in 2003 and declined to 130,461 ha in 2006 (Fig. 2.5), which correlates with the yield variation during this period as reported by Shi and Tao 2014. Rice simulated yield with fertilizer did better capture the long-term observed yield pattern with a RMSE of 154.64 kg/ha, than the simulated yield without fertilizer with a RMSE of 297.42 kg/ha. Its cultivated area increment was reported from 80,312 ha in 2007 to 125,329 ha in 2008. This 1.6 times increase in rice yield after 2008 (Fig. 2.5) is speculated to be attributed to improved management practices and varietal improvements as government policies for rice production changed after 2003 with collective efforts on soil fertility and weed management along with timely harvest (Tollens et al., 2013). Crop yield variability can be attributed either to the expansion of cropland area and/or to the improved cultivar and optimized management practices and were reported by Okuyama et al. (2017), with reference to technical efficiency of production potential of cereal crops in Senegal. The variability in rice yield is far more affected by environmental variables (e.g., rainfall) than the other crops, such as upland rice, whereas lowland rice in humid conditions is relatively less variable (Okuyama et al., 2017).

However, simulated yields of sorghum and millet were consistent with the long-term yield pattern. The simulated yield pattern for sorghum could capture long-term yield trends with RMSE of 135.6 and 87.77 kg/ha for fertilized and no fertilized scenarios, respectively (Fig. 2.5). Similarly, in the case of pearl millet, a RMSE of 127.16 and 36.82 kg/ha for fertilized and no-fertilized scenarios captured long-term yield trends (Fig. 2.5), respectively. Additional differences between the simulated and the observed results could be explained by the fact that simulations did not account for impacts of pest, disease, and weeds.

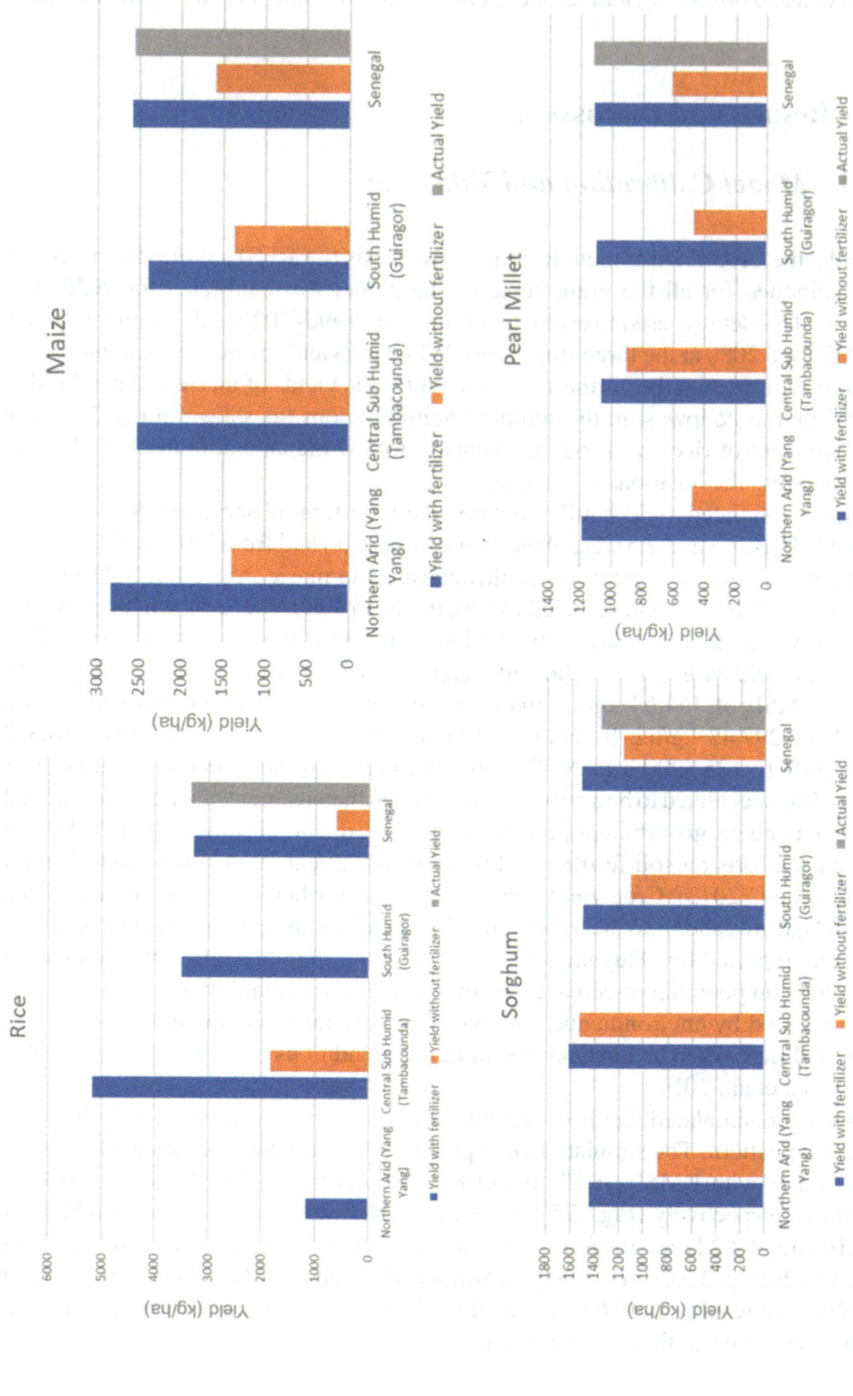

Fig. 2.4 Simulated yield of major cereal crops in Senegal's study sites for 2020. *Source* Author's computations

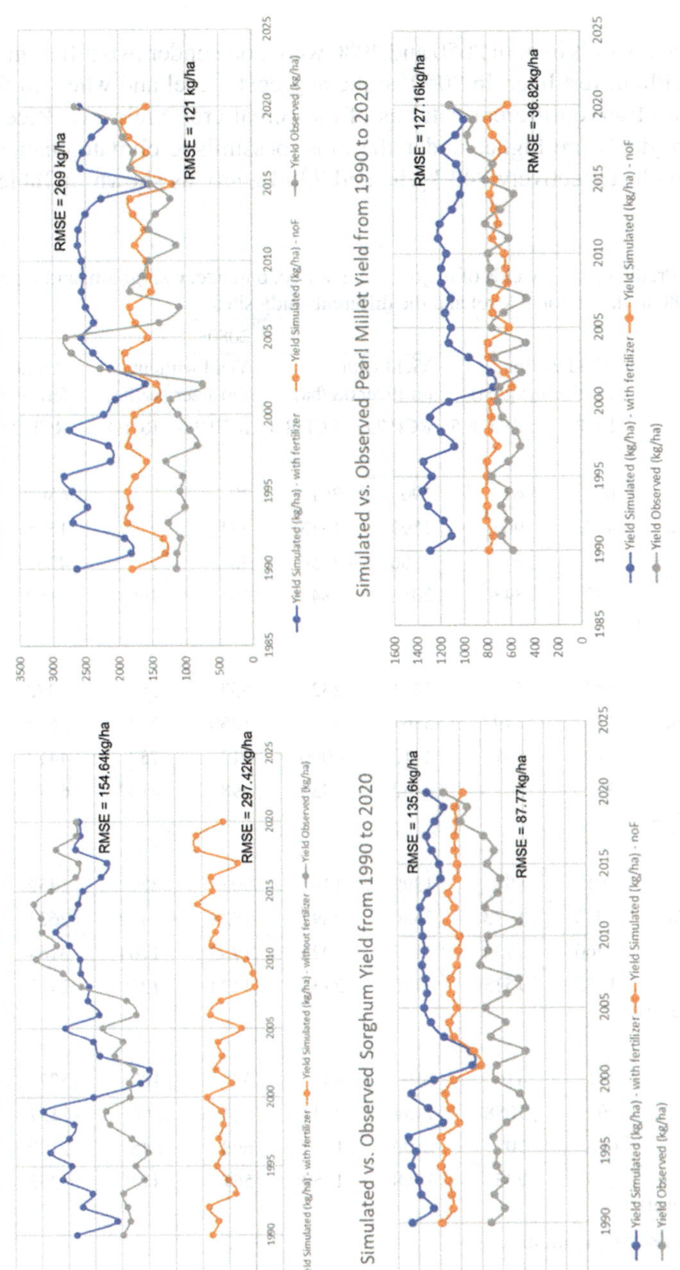

Fig. 2.5 Long-term yield simulations (1990–2020) of major cereal crops: rice, maize, sorghum, and pearl millet for two scenario of No Fertilizer (noF) and Fertilizer (with fertilizer) application. *Source* Author's computations

2.7.2 Simulated Cereal Crop Yields

The simulated yields in both 2050 and 2080 were done under two different scenarios: with and without fertilizer. In 2050, at the aggregate level and when no fertilizer is applied, we observe different patterns of the cereal crop yields. (i) Rice and pearl millet crop yields are higher under the most pessimistic climate change scenario (RCP 8.5 with respectively 343 kg/ha and 735 kg/ha), as shown in Table 2.10. (ii)

Table 2.10 Predicted crop yields of major cereal: **a** rice, **b** maize, **c** sorghum, and **d** pearl millet in 2050 and 2080 at the national level and the different study sites

	2050				2080			
	Yield without fertilizer (kg/ha)		Yield with fertilizer (kg/ha)		Yield without fertilizer (kg/ha)		Yield with fertilizer (kg/ha)	
Region	RCP 2.6	RCP 8.5	RCP 2.6	RCP 8.5	RCP 2.6	RCP 8.5	RCP 2.6	RCP 8.5
Rice								
Yang Yang	66	61	190	361	79	0	656	0
Tambacounda	442	963	1593	1313	371	472	1555	369
Guiragor	38	6	4266	3856	38	7	4266	2231
Senegal (Aggregate level)	182	343	2016	1843	163	160	2159	867
Maize								
Yang Yang	583	528	2271	2327	533	386	2372	892
Tambacounda	1182	1197	5861	5157	1250	772	5657	2840
Guiragor	912	861	4413	4006	912	752	4413	3088
Senegal (Aggregate level)	892	862	4182	3830	898	4147	637	2273
Sorghum								
Yang Yang	901	898	1399	1482	903	821	1452	1080
Tambacounda	1943	1973	2856	2895	1828	1432	2623	1815
Guiragor	1760	1733	3172	2922	1763	1704	3164	2709
Senegal (Aggregate level)	1535	1535	2476	2433	1498	1319	2413	1868
Pearl millet								
Yang Yang	411	416	519	632	380	187	507	233
Tambacounda	699	1088	1454	2131	604	506	1152	629
Guiragor	698	702	2100	1914	698	545	2100	1515
Senegal (Aggregate level)	603	735	1358	1559	561	413	1253	792

Source Author's computations

While sorghum crop yield remains the same in both climate change scenarios (1535 kg/ha), predicted maize crop yield is higher under the most optimistic climate change scenario (RCP 2.6 with 892 kg/ha). Still in 2050, at the aggregate level and with the application of fertilizer, the predicted crop yield for all cereal except pearl millet, the predicted yields are higher under the most optimistic climate change scenario as one would expect (which are respectively 2016 kg/ha, 4182 kg/ha, 2476 kg/ha for rice, maize, and sorghum).

In 2080, at the national level, regardless of the application or not of fertilizer, all the predicted cereal crops yields are higher under RCP 2.6 (which is the most optimistic climate change change) except for maize. Another important observation is that all cereal crop yields seem to have values in the central sub-humid area represented here by Tambacounda. Given these results, we will use the predicted simulated crop yields with the application of fertilizer in order to compute the predicted production in both 2050 and 2080 (Fig. 2.6).

2.7.3 Simulated Production

In addition to the simulated yields, change in planting area was also needed in order to compute the simulated production (Table 2.11 displays the planting area change per crop and year). In 2050, total simulated production varied from 676 to 903.7M kg for rice, 989.2M kg to 1.3B kg for maize, 609M kg to 757 for sorghum, and 1.25B to 1.75B kg for pearl millet (Fig. 2.7 and Table 2.10), depending on the matrix of worst possible scenario (–10% change in planting area and climate change scenario) to best possible scenario (+10% change in planting area and climate change scenario) (see Fig. 2.7 and Table 2.10). Depending on the scenario, the changes in total simulated production of the combined four cereals at country level varied from 3.71 to 4.52B kg (see Fig. 2.8 and Table 2.12).

In 2080, total simulated production varied from 286M kg to 1.1B kg for rice, 148–789M kg for maize, 421–812M kg for sorghum, and 656M kg to 1.55B kg for pearl millet (Fig. 2.7 and Table 2.12), again depending on the matrix of worst possible scenario (–10% change in planting area and climate change scenario) to best possible scenario (+10% change in planting area and climate change scenario) (see Fig. 2.7 and Table 2.12). Depending on the scenario, the changes in total simulated production of the combined four cereals at country level varied from 1.89 to 3.68B kg (see Fig. 2.8 and Table 2.12).

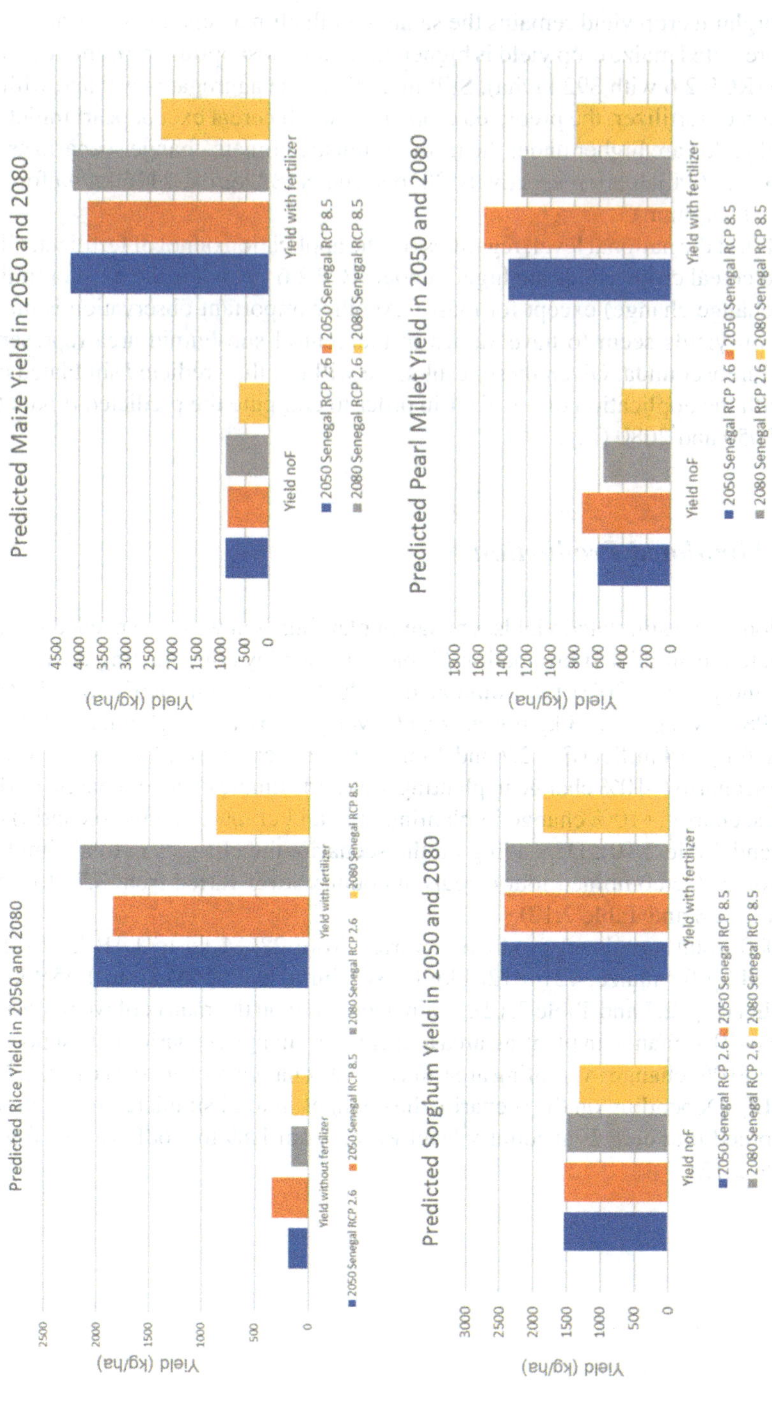

Fig. 2.6 Predicted Major Cereal yields: **a** rice, **b** maize, **c** sorghum, and **d** pearl millet in 2050 and 2080. *Source* Author's computations

Table 2.11 Planting area change per crop and year

Year	2050			
Planting area change (%)	Rice	Maize	Sorghum	Pearl millet
–10	366,774.30	258,279.30	250,290.90	920,758.50
–5	387,150.65	272,628.15	264,195.95	971,911.75
0	407,527.00	286,977.00	278,101.00	1,023,065.00
5	427,903.35	301,325.85	292,006.05	1,074,218.25
10	448,279.70	315,674.70	305,911.10	1,125,371.50
	2080			
Planting area change (%)	Rice	Maize	Sorghum	Pearl millet
–10	330,096.87	232,451.37	225,261.81	828,682.65
–5	367,793.12	258,996.74	250,986.15	923,316.16
0	407,527.00	286,977.00	278,101.00	1,023,065.00
5	449,298.52	316,392.14	306,606.35	1,127,929.16
10	493,107.67	347,242.17	336,502.21	1,237,908.65

Source Author's computations

Based on the simulation results, there is going to be a decrease in crop production from 2050 to 2080 which varies from 1.82 to 8.37 B kg. Additionally, since cereal production in 2020 was 3.6 B kg (FAO, 2022), there is a projected increase in crop production ranging from 75.3 to 88.7M kg in 2050. However, comparing the cereal production in 2020 with the simulated results in 2080, one can see that, in the worst case scenario, there is a projected decrease in cereal production of 1.74B kg. Whereas, in the best case scenario, there is a projected increase in cereal production of 49.4M kg.

Another interesting finding is about the production trend. In 2022, based on FAOSTAT data, rice was the main produced cereal with 1.3 M kg followed by pearl millet (1.1M kg) maize (762,000 kg), and sorghum(377,000 kg) (see Appendix A.3). However, based on the simulated production results in 2050 and 2080, pearl millet becomes the main produced crop followed respectively by maize, rice, and sorghum, under all scenarios except the simulated results obtained in 2080 under RCP 2.6 climate change scenario (see Fig. 2.9).

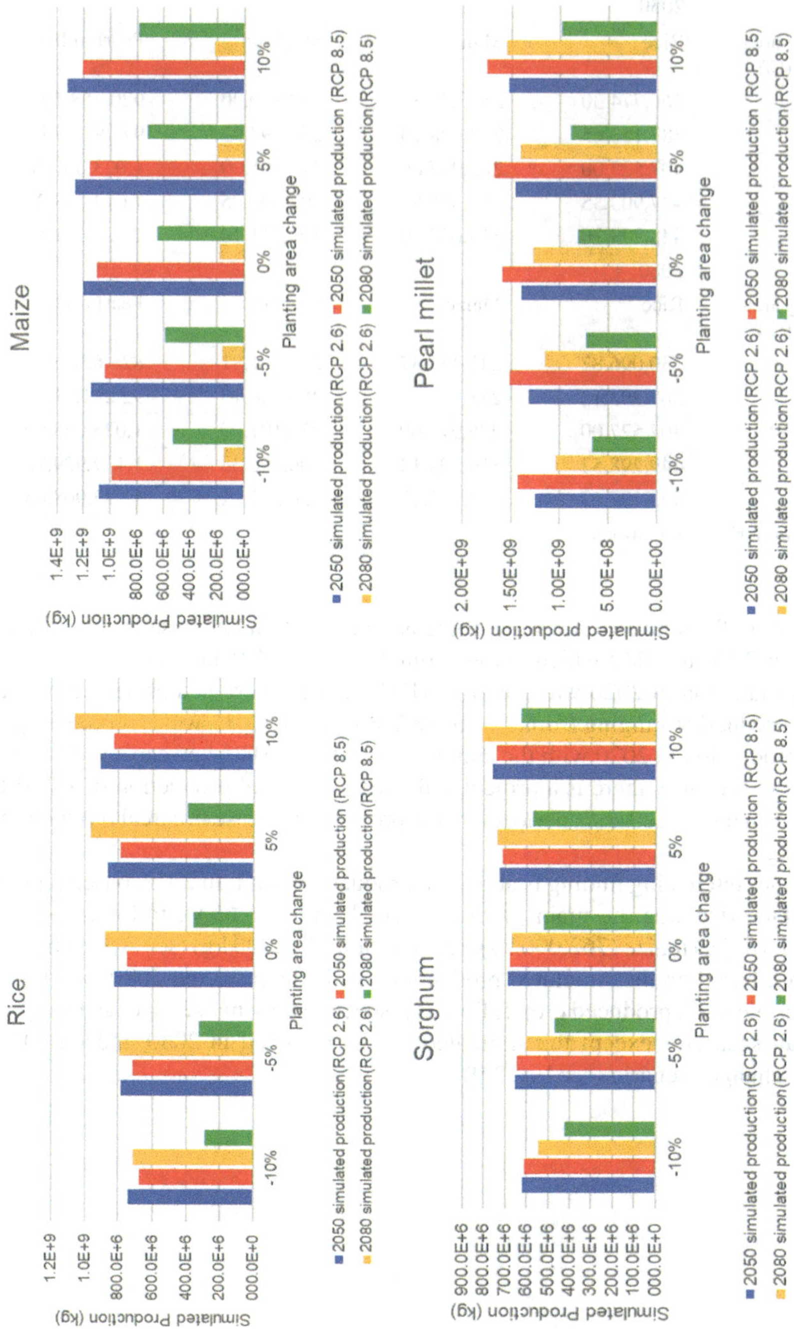

Fig. 2.7 Simulated Major Cereal crop production in 2050 and 2080. *Source* Author's computations

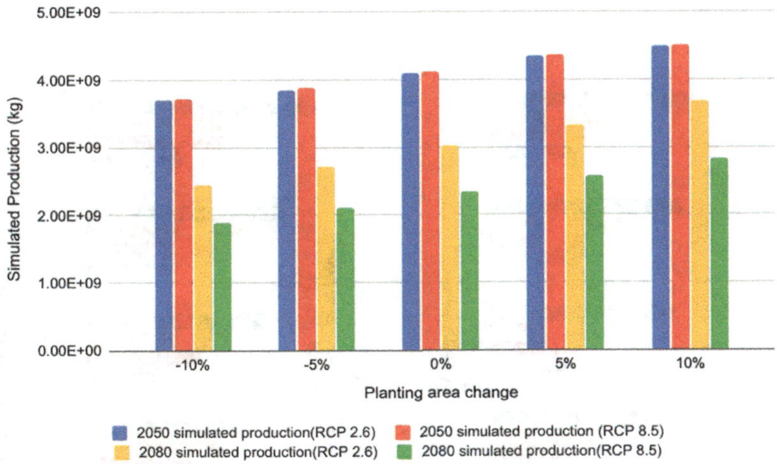

Fig. 2.8 Estimate aggregate cereal production for 2050 and 2080. *Source* Author's computations

Table 2.12 Estimate value of major cereal crops: rice, maize, sorghum, and pearl millet for 2050 and 2080 in Senegal

	2050		2080	
	Simulated production (kg)		Simulated production (kg)	
Planting area change	RCP 2.6	RCP 8.5	RCP 2.6	RCP 8.5
Rice				
–10%	7.39E+08	6.76E+08	7.13E+08	2.86E+08
–5%	7.81E+08	7.14E+08	7.94E+08	3.19E+08
0%	8.22E+08	7.51E+08	8.80E+08	3.53E+08
5%	8.63E+08	7.89E+08	9.70E+08	3.90E+08
10%	9.04E+08	8.26E+08	1.10E+09	4.28E+08
Maize				
–10%	1.10E+09	9.89E+08	1.48E+08	5.28E+08
–5%	1.10E+09	1.00E+09	1.65E+08	5.89E+08
0%	1.20E+09	1.10E+09	1.83E+08	6.52E+08
5%	1.30E+09	1.20E+09	2.02E+08	7.19E+08
10%	1.30E+09	1.20E+09	2.21E+08	7.89E+08
Sorghum				
–10%	6.20E+08	6.09E+08	5.44E+08	4.21E+08
–5%	6.54E+08	6.43E+08	6.06E+08	4.69E+08
0%	6.89E+08	6.77E+08	6.71E+08	5.20E+08
5%	7.23E+08	7.11E+08	7.40E+08	5.73E+08
10%	7.57E+08	7.44E+08	8.12E+08	6.29E+08
Pearl millet				
–10%	1.25E+09	1.44E+09	1.04E+09	6.56E+08
–5%	1.32E+09	1.52E+09	1.16E+09	7.31E+08
0%	1.39E+09	1.59E+09	1.28E+09	8.10E+08
5%	1.46E+09	1.67E+09	1.41E+09	8.93E+08
10%	1.53E+09	1.75E+09	1.55E+09	9.80E+08

Source Author's computations

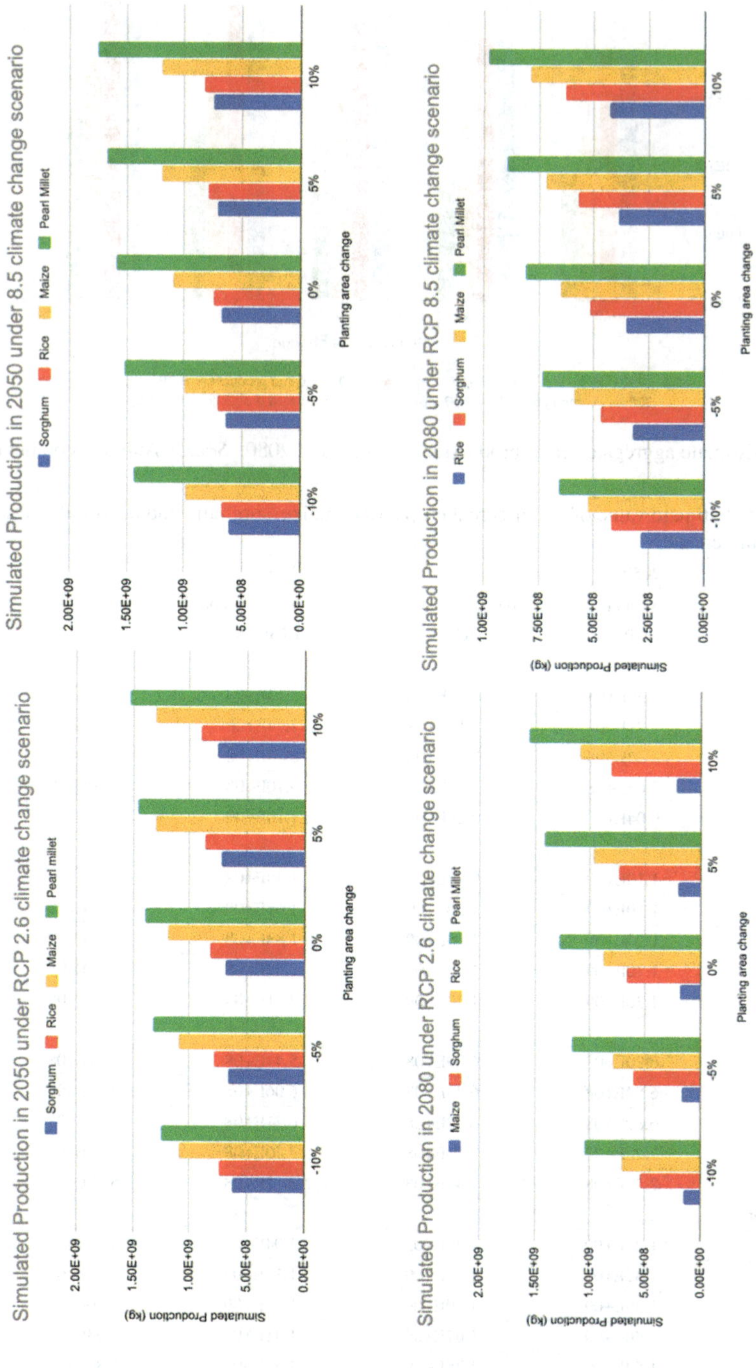

Fig. 2.9 Simulated production of major crops in 2050 and 2080 based on planting area change. *Source* Author's computations

References

Aboelkhair, H., Morsy, M., & El Afandi, G. (2019). Assessment of agroclimatology NASA POWER reanalysis datasets for temperature types and relative humidity at 2 m against ground observations over Egypt. *Advances in Space Research, 64*(1), 129–142. [35]

Amouzou, K. A., Naab, J. B., Lamers, J. P., & Becker, M. (2018). CERES-maize and CERES-Sorghum for modeling growth, nitrogen and phosphorus uptake, and soil moisture dynamics in the dry savanna of West Africa. *Field Crop Research, 217*, 134–149. [44]

Araya, A., Kisekka, I., Gowda, P. H., & Prasad, P. V. V. (2017). Evaluation of water-limited cropping systems in a semi-arid climate using DSSAT-CSM. *Agricultural Systems, 150*, 86–98. [47]

Bai, J., Chen, X., Dobermann, A., Yang, H., Cassman, K. G., & Zhang, F. (2010). Evaluation of NASA satellite-and model-derived weather data for simulation of maize yield potential in China. *Agronomy Journal, 102*(1), 9–16. [33]

Balde, R., Boly, M., & Avenyo, E. K. (2020). Labour market effects of COVID-19 in sub-Saharan Africa: An informality lens from Burkina Faso, Mali, and Senegal. In *Maastricht Economic and Social Research Institute on Innovation and Technology (UNU-MERIT)*. Maastricht. [62]

de Vries, M. E., Sow, A., Bado, V. B., & Sakane, N. (2012). Simulation of potential yields of new rice varieties in the Senegal River Valley. In *Improving Soil Fertility Recommendations in Africa using the Decision Support System for Agrotechnology Transfer (DSSAT)* (pp. 141–155). Springer, Dordrecht. [43]

Ekeleme, F., Kamara, A., Omoigui, L. O., Tegbaru, A., Mshelia, J., & Onyibe, J. (2008). Guide to Rice Production in Borno State, Nigeria. International Institute of Tropical Agriculture (IITA). [54]

Elias, A. A., Robbins, K. R., Doerge, R. W., & Tuinstra, M. R. (2016). Half a century of studying genotype × environment interactions in plant breeding experiments. *Crop Science, 56*(5), 2090–2105. [50]

FAO. (2012). FAO: Food security data and definitions. http://www.fao.org/economic/ess/ess-fs/fs-data/ess-fadata/en/ [27]

FAO (Food and Agriculture Organization of the United Nations). (2022). FAOSTAT database. Retrieved May 15, 2020 from. http://www.fao.org/faostat/en/#data. [32]

Fall, S., Semazzi, F. H. M., Niyogi, D. D. S., Anyah, R. O., & Bowden, J. (2006). The spatiotemporal climate variability over Senegal and its relationship to global climate. *International Journal of Climate, 26*, 2057–2076. [24]

Grassini, P., van Bussel, L. G., Van Wart, J., Wolf, J., Claessens, L., Yang, H., & Cassman, K. G. (2015). How good is good enough? Data requirements for reliable crop yield simulations and yield-gap analysis. *Field Crop Research, 177*, 49–63. [30]

Han, E., Ines, A. V., & Koo, J. (2019). Development of a 10 km resolution global soil profile dataset for crop modeling applications. *Environmental Modelling & Software, 119*, 70–83. [40]

Hengl, T., de Jesus, J. M., MacMillan, R. A., Batjes, N. H., Heuvelink, G. B., Ribeiro, E., Samuel-Rosa, A., Kempen, B., Leenaars, J. G., Walsh, M. G., & Gonzalez, M. R. (2014). SoilGrids1 km—Global soil information based on automated mapping. *PloS One, 9*(8), e105992. [41]

Hijmans, R., et al. (2005). Very high resolution interpolated climate surfaces of global land areas. *International Journal of Climatology,25*, 1965–1978. https://doi.org/10.1002/joc.1276

Hoogenboom, G., Porter, C. H., Boote, K. J., Shelia, V., Wilkens, P. W., Singh, U., White, J. W., Asseng, S., Lizaso, J. I., Moreno, L. P., Pavan, W., Ogoshi, R., Hunt, L. A., Tsuji, G. Y., & Jones, J. W. (2019). *The DSSAT crop modeling ecosystem* (pp. 173–216).; Boote, K. J. (Ed.). *Advances in crop modeling for a sustainable Agriculture*. Burleigh Dodds Science Publishing. https://doi.org/10.19103/AS.2019.0061.10 [61]

Hoogenboom, G., Porter, C. H., Shelia, V., Boote, K. J., Singh, U., White, J. W., Hunt, L. A., Ogoshi, R., Lizaso, J. I., Koo, J., Asseng, S., Singels, A., Moreno, L. P., Jones, J. W. (2017). Decision support system for Agrotechnology transfer (DSSAT) version 4.7 (https://DSSAT.net). DSSAT Foundation. [38]

Ilboudo, A., Soulama, S., Hien, E., & Zombre, P. (2020). Peasant perceptions of the degradation of natural resources in lowlands in the Sudano-Sahelian zone: Case of the Nakanbé-Dem watershed in Burkina Faso. *International Journal of Chemical and Biological Sciences, 14*(3), 883–895. [37]

Jha, P. K., Araya, A., Stewart, Z. P., Faye, A., Traore, H., Middendorf, B. J., & Prasad, P. V. V. (2021). Projecting potential impact of COVID-19 on major cereal crops in Senegal and Burkina Faso using crop simulation models. *Agricultural Systems,190*, 103107. https://doi.org/10.1016/j.agsy.2021.103107 [28]

Jha, P. K., Ines, A. V., & Singh, M. P. (2021). A multiple and ensembling approach for calibration and evaluation of genetic coefficients of CERES-maize to simulate maize phenology and yield in Michigan. *Environmental Modelling & Software,* 104901. [49]

Jha, P. K., Kumar, S. N., & Ines, A. V. (2018). Responses of soybean to water stress and supplemental irrigation in upper Indo-Gangetic plain: Field experiment and modeling approach. *Field Crop Research, 219*, 76–86. [48]

Jones, P. G., & Thornton, P. K. (2000). MarkSim: Software to generate daily weather data for Latin America and Africa. *Agronomy Journal,9*, 445–453. https://doi.org/10.2134/agronj2000.923445x [63]

Jones, J. W., Hoogenboom, G., Porter, C. H., Boote, K. J., Batchelor, W. D., Hunt, L. A., et al. (2003). The DSSAT cropping system model. *European Journal of Agronomy, 18*(3), 235–265. https://doi.org/10.1016/S1161-0301(02)00107-7 [58]

Jones, P. G., Thornton, Philip, Díaz, W., & Wilkens, P. W. (2002). *MarkSim: A computer based tool that generates simulated weather data for crop modeling and risk assessment*. CD-ROM Series, CIAT. https://doi.org/10.3109/9780203427583-32 [64]

Khairwal, I. S., Rai, K. N., Diwakar, B., Sharma, Y. K., Rajpurohit, B. S., & Nirwan, B. Bhattacharjee, R. (2007). Pearl millet: Crop management and seed production manual. In *International Crops Research Institute for the Semi-Arid Tropics (ICRISAT)*. Patancheru. [53]

Koo, J., & Dimes, J. (2010). HC27: Generic/Prototypical Soil Profiles. International Food Policy Research Institute. University of Minnesota. http://harvestchoice.org/node/2239. [42]

Leippert, F., Darmaun, M., Bernoux, M., & Mpheshea, M. (2020). *The potential of agroecology to build climate-resilient livelihoods and food systems*. Food and Agricultural Organization and Biovision. [29]

Lobell, D. B., Cassman, K. G., & Field, C. B. (2009). Crop yield gaps: Their importance, magnitudes, and causes. *Annual Review of Environment and Resources, 34*, 179–204. [52]

Lobell, D. B., Ortiz-Monasterio, J. I., & Lee, A. S. (2010). Satellite evidence for yield growth opportunities in Northwest India. *Field Crop Research, 118*(1), 13–20. [34]

Maguèye, M., & Ndao, M. (2013). Climate risk and food security in Senegal: Analysis of climate impacts on food security and livelihoods. [19]

McSweeney, C., New, M., Lizcano, G., & Lu, X. (2010) The UNDP climate change country profiles: Improving the accessibility of observed and projected climate information for studies of climate change in developing countries. *Bulletin of the American Meteorological Society, 91*(2), 157–166. [25]

Okuyama, Y., Maruyama, A., Takagaki, M., & Kikuchi, M. (2017). Technical efficiency and production potential of selected cereal crops in Senegal. *Journal of Agriculture and Rural Development in the Tropics and Subtropics (JARTS), 118*(2), 187–197. [56]

Ouedraogo, I., & Ayantunde, A. A. (2019). Biomass production and management practices in a mixed-crop-livestock system in the West African Sahel: Opportunities and constraints. Report. International Livestock Research Institute. [67]

Parkes, B., Higginbottom, T. P., Hufkens, K., Ceballos, F., Kramer, B., & Foster, T. (2019). Weather dataset choice introduces uncertainty to estimates of crop yield responses to climate variability and change. *Environmental Research Letters, 14*(12), 124089. [36]

Pickering, N. B., Hansen, J. W., Jones, J. W., Wells, C. M., Chan, V. K., & Godwin, D. C. (1994). Weatherman: A utility for managing and generating daily weather data. *Agronomy Journal, 86*(2), 332–337. [39]

Rware, H., Wairegi, L., Oduor, G., Macharia, M., Romney, D., Tarfa, B. D., de Maria, R., Ley, G., Tetteh, F., Makumba, W., & Dicko, M. (2014). Assessing the potential to change stakeholder's knowledge and practices on fertilizer recommendations in Africa. *Agricultural science, 5*(14), 1384. [69]

Saito, K., Vandamme, E., Johnson, J. M., Tanaka, A., Senthilkumar, K., Dieng, I., & Lamare, D. (2019). Yield-limiting macronutrients for rice in sub-Saharan Africa. *Geoderma, 338*, 546–554. [66]

Sane, Y., Panthou, G., Bodian, A., Vischel, T., Lebel, T., Dacosta, H., Quantin, G., Wilcox, C., Ndiaye, O., Diongue-Niang, A., & Diop Kane, M. (2018). Intensity—Duration—Frequency (IDF) rainfall curves in Senegal. *Natural Hazards and Earth System Sciences (NHESS), 18*(7), 1849–1866. [57]

Shi, W., & Tao, F. (2014). Vulnerability of African maize yield to climate change and variability during 1961–2010. *Food Security, 6*(4), 471–481. [71]

Singh, P., Boote, K.J., Kadiyala, M. D. M., Nedumaran, S., Gupta, S. K., Srinivas, K., & Bantilan, M. C. S. (2017). An assessment of yield gains under climate change due to genetic modification of pearl millet. *Science of the Total Environment, 601*, 1226–1237. [46]

Singh, P., Nedumaran, S., Traore, P. C. S., Boote, K. J., Rattunde, H. F. W., Prasad, P. V. V., Singh, N. P., Srinivas, K., & Bantilan, M. C. S. (2014). Quantifying potential benefits of drought and heat tolerance in rainy season sorghum for adapting to climate change. *Agricultural and Forest Meteorology, 185*, 37–48. [45]

Sommer, R., Bossio, D., Desta, L., Dimes, J., Kihara, J., Koala, S., Mango, N., Rodriguez, D., Thierfelder, C., & Winowiecki, L. (2013). Profitable and Sustainable Nutrient Management Systems for East and Southern African Smallholder Farming Systems Challenges and Opportunities: A Synthesis of the Eastern and Southern Africa Situation in Terms of Past Experiences, Present and Future Opportunities in Promoting Nutrients Use in Africa (p. 91). CIAT. [55]

Spinage, C. A. (2012). The Changing Climate of Africa Part II: West Africa and the Sahel. In C. A. Spinage (Ed.), *African ecology—Benchmarks and historical perspectives*. Springer. [26]

Stewart, Z. P., Pierzynski, G. M., Middendorf, B. J., & Prasad, P. V. V. (2020). Approaches to improve soil fertility in sub-Saharan Africa. *Journal of Experimental Botany, 71*(2), 632–641. [65]

Tollens, E., Demont, M., Sié, M., Diagne, A., Saito, K., Wopereis, M. C., Johnson, D. E., Ahmadi, N., & Jalloh, A. (2013). From WARDA to AfricaRice: An overview of rice research for development activities conducted in partnership in Africa. In *Realizing Africa's Rice Promise 1–23*. CABI, Wallingford. [72]

Van Wart, J., Grassini, P., Yang, H., Claessens, L., Jarvis, A., & Cassman, K. G. (2015). Creating long-term weather data from thin air for crop simulation modeling. *Agricultural and Forest Meteorology, 209*, 49–58. [31]

Walker, T. S., Alwang, J., Alene, A., Ndjuenga, J., Labarta, R., Yigezu, Y., Diagne, A., Andrade, R., Andriatsitohaina, R. M., De Groote, H., & Mausch, K. (2015). Chapter 19 varietal adoption, outcomes, and impact. In T. S. Walker & J. Alwang (Eds.), *Crop improvement, adoption, and impact of improved varieties in food crops in sub-Saharan Africa* (p. 388). [68]

Willmott, C. J. (1982). Some comments on the evaluation of model performance. *The Bulletin of the American Meteorological Society (BAMS), 63*(11), 1309–1313 [70]

World Food Programme. (2012). Food Security Monitoring System (SSSAS—Système de suivi de la sécurité alimentaire au Sénégal). WFP. [15]

Worou, O. N., Tondoh, J. E., Sanou, J., Gaiser, T., Nikiema, P. M., Bayala, J., Bazié, P., Ky-Dembele, C., & Kalinganiré, A. (2018). Intensifying maize production under climate change scenarios in central West Burkina Faso. In W. L. Filho (Ed.), *Handbook of Climate Change Resilience* (pp. 1–23). Springer International Publishing.

Chapter 3
Alternatives and Robustness Analysis

3.1 Introduction

Since agricultural production is inherently variable, the immediate step after our prediction analysis is to explore the impact of technical changes, alternative soil quality, and longer persistent disasters on cereal production. As discussed in the first chapter, the most common climate event risks to agricultural production in Senegal are droughts and floods. The incidence of these and other adverse events is indicated in Fig. 3.1, largely based on reports of adverse events for the period 1973–2020. During the 48-year period 1973–2020, Senegal's agricultural sector has been subjected to several major shocks. Erratic rainfall and drought emerge as the most common sources of production shocks, followed by locusts. Related risk events may occur in isolation, but can also present as multiple, overlapping shocks—as was the case in 1984, 2002, 2007, and 2011—with far greater impacts and higher associated losses. Those losses account for approximately 50% of crop yield reductions. Additionally, in West Africa, an increase in the frequency and the magnitude of extreme events (heavy rains, heat waves) is expected as well as a decline and a slight increase in annual rainfall in the west and east of the region, respectively (Diedhiou et al., 2018). Rainfed crop production dominates in West Africa, making the region a climate change hotspot.

Moreover, Senegalese agriculture is subject to losses exceeding 10% of gross production value in one out of every five or six years on average due to unmanaged risk. Measured in terms of gross agricultural value, crop production in Senegal was significantly reduced 11 times by adverse events during the period 1980–2012, for an overall frequency of one in three to four years on average (FAO, 2023). All but three of these events resulted in a drop in aggregate production value of 10–30%. It is worth noting that three (1980, 1983, 2002) of the four years with the highest crop losses coincided with extreme drought events, whereas poor rainfall affecting key crop production areas coupled with locust and bird infestation contributed to high crop losses in 2007 and 2011.

M. Toure, *Agricultural Adaptation to Climate Change in Sub-Saharan Africa*,
Advances in African Economic, Social and Political Development,
https://doi.org/10.1007/978-3-031-80716-9_3

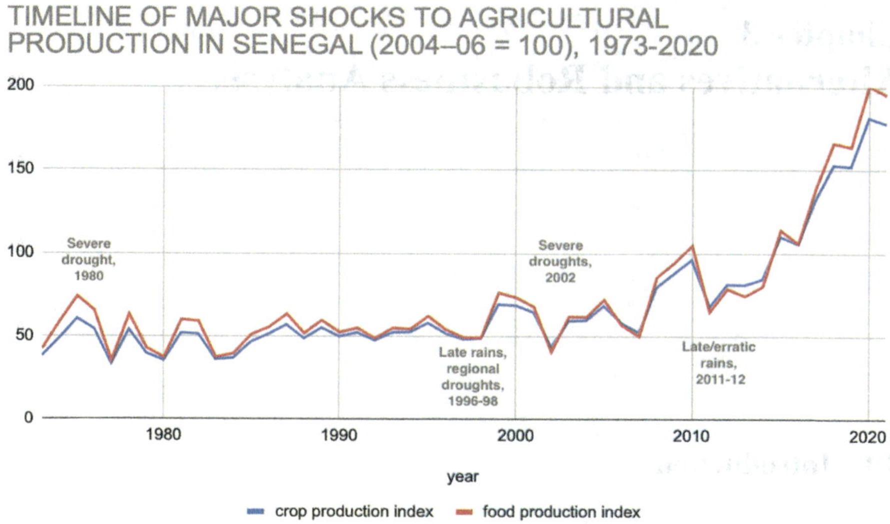

Fig. 3.1 Timeline of major shocks to Agricultural production in Senegal from 1973 to 2020. *Source* "Senegal Agricultural Sector Risk Assessment" by DataBank | World Development Indicators 2023

The main contribution of this chapter is to investigate the role of technology, changing quality of soil, and different scenarios of changing arable land (due to potential persistent natural disaster events) in future crop production. Each of these factors can either positively or negatively affect crop yield and/or available planting area. Section 3.2 discusses the potential long-run impacts of promising technologies in crop yields. For that, we evaluate a selected promising and alternative technologies for crops, specifically drought- and heat-tolerant varieties for rice, maize, pearl millet, and sorghum, which have been used in Senegal and sub-Saharan Africa in general. In Sect. 3.3, present and predicted soil properties and their predicted impact on long-term cereal production are discussed. After that, Sect. 3.4 focuses on planting area change scenarios as a result of potential natural disaster events and the ways that could cause the variation of crop production simulation results that were obtained in Chap. 2. This chapter ends with an analysis of the overall predicted variations in crop production.

3.2 Technical Changes

A recent study, using crop simulation models to examine the effects of abiotic stresses (temperature and precipitation), estimated that the long-term changes in temperature and rainfall due to climate change could reduce global maize, rice, sorghum, and pearl millet yields in 2050 by as much as 25% compared with 2010 yields in the absence of technological change and market effects (Rosegrant et al. 2005, 2014). Clearly, new technologies will be developed and economic agents will respond to changes in prices over time due to changes in population, income, and climate. Nevertheless, significant increases in prices may occur. Higher prices would particularly hurt marginalized populations, as well as those who spend a large share of their income on food. The authors found that a range of technology improvements could help reduce some of these adverse impacts.

Other recent work has examined the impact of climate change on crop yields, production, and prices in 2050 relative to a baseline without climate change. Drawing on multiple climate, crop, and economic models, Hachigonta et al. 2013 (Hachigonta et al., 2013) estimated that yields of four major crops (maize, rice, wheat, and soybeans) in 2050 would be around 11% lower due to the effects of climate change once economic responses were considered. Together, these studies indicate the importance of improved modeling to disentangle the complex interactions between multiple drivers of change. Improved understanding of emerging technologies must be used to explore alternative strategies to achieve and sustain increases in agricultural productivity and food security. Since basic and applied research in agronomy, soil science, ecology, and entomology has generated a wealth of technologies and practices to increase the resilience of agroecosystems and to protect and increase productivity, the following section will be evaluating a selection of promising and alternative technologies for crops, specifically drought- and heat-tolerant varieties. For that, we explore different crop cultivars, which are plant varieties that scientists and agronomists have selected and tested for desired traits of improved yield in sub-Saharan Africa.

3.2.1 New Promising and Alternative Varieties of Cereal Crops

In the previous chapter, long-term simulations were performed using the same cultivars irrespective of decadal change of varietal improvement (Walker et al., 2015). In this chapter, in order to simulate technical changes, we will perform long-term simulations using different cultivars (based on further studies done to improve crop yield in sub-Saharan Africa) for each crop: rice, sorghum, maize, and pearl millet.

3.2.1.1 Rice

Rice is the world's most important food crop and is estimated as the staple food for half of humanity. The demand for rice is expected to increase with population growth. West African countries like Senegal, Guinea-Bissau, and Guinea are highly dependent on rice for food calories. According to the International Food Policy Research Institute (Glatzel, 2018), the average Senegalese consumes about 85 kg of rice each year. Despite the strong political commitment to improve rice production, inadequate

Table 3.1 Characteristics of the cultivars used in the experiments in Senegal

Short name	Variety	Genetic group	Cycle length (days)
NE4	Nerica 4	Hybrid	95–100
DK3	Dkap 3	Hybrid	N.A.

Source "Adapting rainfed rice to climate change: a case study in Senegal" by Gérardeaux et al. (2021)

water, soil, and crop management still compromises paddy yield productivity. Rice production is also threatened by variability in annual rainfall, fluctuations in the onset of growing season, and increase in aridity (Koudahe et al., 2017). Assessments of the changes in local climatic variables and their impact on rice yield are currently scarce.

Since the scope for increase in rice production under irrigated conditions is uncertain, rainfed rice is therefore a key component for regional food security impelling agronomists to assess climate change impact on rainfed rice yield and to design rainfed rice ideotypes suited to future climate scenarios. While we are using the same DSSAT CSM-CERES-Rice model as in Chap. 2, we are going to explore two breeding cultivars with the same calibrated characteristics that (Gérardeaux et al., 2021) used to examine the potential impact of climate change on rainfed rice yield in Senegal (see Table 3.1). Their calibrated cultivar parameters used as proxy for physiological traits investigated the interaction between genotype and environment, and designed rice ideotypes for adaptation to climate in Senegal. Those breeding cultivars are expected to offset yield losses by limiting rice sterility with high temperatures (Table 3.2).

The rice team (i.e., a partnership between the International Fertilizer Development Center (IFDC), University of Florida, and the global Agricultural Model Intercomparison and Improvement Project (AgMIP)) has studied rice models under varying climatic conditions (Hasegawa et al., 2017) and proposed a new set of cultivar dependent parameters, the so-called genetic coefficients, to ensure that CERES-Rice correctly simulated the effect of high temperatures, drought, and increases in atmospheric [CO_2]. These coefficients control crop development, such as flowering and maturing. There are eleven genetic coefficients defined in the CERES-Rice model and Table 3.3 provides the values for the two selected cultivars: DK3 and NE4. THOT, TCLDP, and TCLDS are the new temperature coefficients used to improve crop response to extreme temperatures. Plant growth, development, and yield depend on species- and cultivar-dependent parameters determining the sensitivity of crops to variations in their environment. A brief description of the cultivars' genetic coefficients can be found in Appendix A.2.

Table 3.2 Values of the calibrated genetic parameters (CERES-Rice model)

Genetic coefficients (unit)	DK3	NE4
P1 (Growing degree days (GDD) (°C))	417.4	417.6
P20 (hours)	19.9	18.8
P2R (Growing degree days (GDD) (°C))	345.2	327.6
P5 (Growing degree days (GDD) (°C))	10.2	10
G1	48.6	70.5
G2 (mg)	0.0326	0.0376
G3	0.64	0.99
PHINT	83	83
THOT	27.8	27.7
TCLDP	15	15
TCLDS	15	15

Source "Adapting rainfed rice to climate change: a case study in Senegal" by Gérardeaux et al. (2021)

Table 3.3 Values of the calibrated genetic parameters (CERES-Maize model)

Genetic coefficients (unit)	TZE COMP4 C2	TZB-SR
P1 (°C/day)	210	290
P2 (day)	0.1	0.6
P5 (°C/day)	660	775
G2 (grains/ear)	850	990
G3 (mg/day)	9.7	6.8
PHINT (°C/day)	55	45

Source "Simulating planting date and cultivar e ects on dryland maize production using CERES-maize model" by Jibrin (2012)

3.2.1.2 Maize

Maize is the major food, feed, and industrial crop globally, and the leading staple food crop in many developing countries. The area coverage, production, and yield of maize have increased over the last 50 years, and much of the increase was recorded in the developing world. Despite these improvements, demand for maize has been outpacing production, partly due to growing demand (for food, feed, energy, and other industries). The demand for maize is expected to double by 2050 (Robinson et al., 2015), which signals the need to drastically increase productivity. Yet maize production and productivity is constrained by several biotic and abiotic factors whose magnitudes are expected to increase under climate change. The major abiotic constraints are recurrent drought, low soil fertility, high soil acidity, soil erosion, heat stress, and waterlogging. Climate change is exacerbating the existing problems and posing new challenges. Moreover, most farmers in Africa rely on rainfall to grow maize; hence, production is very vulnerable to climate shocks, and dry conditions can have disastrous consequences.

In recent years, evidence has emerged demonstrating the negative effect of high temperatures on the performance of maize. Research in Africa south of the Sahara has shown that a decrease in rainfall by 20% is less detrimental to maize yields than a 2 °C increase in temperatures (Lobell & Burke, 2010). Elevated temperatures impact maize yields by shortening the life cycle, reducing light interception, and increasing sterility (Cairns et al., 2012; Stone, 2001). In order to assess the promising improvement of maize yield due to technical changes, we will explore two open-pollinated maize cultivars: TZB-SR and TZE COMP4, which have been used by Jibrin (2012) whose objective was to predict long-term cultivar effects on maize production sub-Saharan Africa. TZB-SR is a widely used late maturing cultivar (120 days to maturity), while TZE COMP4 C2 is an early-maturing cultivar (90 days to maturity).

After evaluating the role of physical characteristics of the seed on the stability of resistance of maize varieties to maize weevil, Akpodiete & Lale (2015) found that TZB-SR and TZE COMP4 cultivars are the most resistant of the varieties tested. They also argue that these two cultivars could be cultivated by farmers to reduce pest damage in storage and could serve as a basis for extension services to farmers and possibly a reference for policy making in the agricultural sector. Those two cultivars are still going to be applied to the same CERES-Maize model that was used in Chap. 2 (see Table 3.3 for cultivars's genetic coefficients). A brief description of the cultivars' genetic coefficients can be found in Appendix A.2.

3.2.1.3 Sorghum

Sorghum (Sorghum bicolor) is a high-yielding, C4 cereal crop, adapted to hot and semi-arid climates (Taylor, 2015). It demonstrates a high level of phenotypic adaptation mechanisms to drought stress, such as leaf rolling to reduce water loss from stomata (Akinseye et al., 2017; Gano et al., 2021). Sorghum is a staple crop to 500 million people worldwide, making it the fifth most important cereal crop globally (Akinseye et al., 2017). In West Africa, the total cultivated land is 28.42 million ha, producing an estimated 28.61 million tons annually (Gano et al., 2021). Sorghum is considered as a sustenance crop for many small-holder farmers in the region and it is frequently observed as the only viable grain crop for farmers in the region (Taylor, 2015).

However, sorghum production is highly variable between growing seasons as West Africa is characterized by seasonal drought and high inter-annual variation in precipitation patterns. The region is furthermore sensitive to effects of climate change (Gano et al., 2021; Haussman et al., 2012). For example, Sultan et al. (2013) predicted up to a 20% loss in yield and higher overall variability due to reduced rainfall. The dependency of sorghum as a staple crop in West Africa is expected to be increasing in the period 2000–2050 and small-holder farmers with rainfed systems

Table 3.4 Values of the calibrated genetic parameters (CERES-Sorghum model)

Genetic coefficients (unit)	CSM335	Fadda
P1 (°C/day)	450	420
P2 (°C/day)	142	102
P3 (°C/day)	202.5	152.5
P4 (°C/day)	81.5	81.5
P5 (°C/day)	440	500
P20 (hour)	12.6	13.2
PANTH (°C/day)	617.5	617.5
Thermal time-sum (°C/day)	1933.5	1873.5

Source "Assessing crop model improvements through comparison of sorghum (sorghum bicolor L. moench) simulation models: A case study of West African varieties" by Akinseye et al. (2017)

are the most vulnerable to variations in rainfall (Haussman et al., 2012). Thus, to be able to improve yield stability in the region, crop breeding efforts must target improved crop cultivars with higher tolerance or adaptability to drought stress. But similarly to crop yield, drought tolerance is a polygenic trait, controlled by numerous genes, whose expressions are sensitive to interactions between the genotype and environment (Parent & Tardieu, 2014).

The varieties CSM335 and Fadda were selected in this study for their contrasting phenology and morphology as well as their responses to assessing crop model improvements through comparison of sorghum(sorghum bicolor L. moench) simulation models, just as Akinseye et al. (2017) used. Variety CSM335 otherwise called 'left Tieble' is a traditional local variety with medium physiological maturity ranging from 105 to 135 days, intermediate plant height, high biomass, low grain and moderate PPS. Variety Fadda is an improved hybrid, medium maturity days (100–135), high-yielding dual purposes (biomass and grain) (see Table 3.4 for cultivars's genetic coefficients). A brief description of the cultivars' genetic coefficients can be found in Appendix A.2.

3.2.1.4 Pearl Millet

The productivity of the Sahelian millet-based systems is still low despite the great wealth of research trials carried out to alleviate the soil fertility and climate constraints over the last 30 years (Akponikpe et al., 2010). After studying the effect of photoperiod and temperature for seven Sahelian millet cultivars in Niger, they found that late sowing had a detrimental effect on most agronomic state variables (leaf number, LAI, stover and grain yield and components) but the extent of this effect was cultivar dependent. Photoperiod effect was observed only on one cultivar (the landrace Maewa). Temperature played a major role in the effect of sowing date on millet development and yield both in terms of amplitude (and independently of photoperiod) and accumulated degree days. Their study provided genetic and

Table 3.5 Values of the calibrated genetic parameters (CERES-Pearl millet model)

Genetic coefficients (unit)	Zatib	Heine Kirey
P1 (°C/day)	340	396
P20 (hour)	12	12
P2R (°C/day)	260	410
P5 (°C/day)	285	285
G1	0.45	0.6
G4	0.9	0.8
PHINT (°C/day)	43	43
GT	1	1.2
G5 (mg)	11	11

Source "Role of Physical Characteristics of the Seed on the Stability of Resistance of Maize Varieties to MaizeWeevil (Sitophilus Zeamais Motschulsky)" by Akponikpe et al. (2010)

eco-physical parameters for the two Sahelian millet cultivars that we are going to use: Zatib and Heine Kirey. While Zatib is an improved cultivar from a breeding selection from Zanfarwa and Tchin-Bijini ecotypes from Niger, Heine Kirey is a domesticated, locally adapted, traditional variety of cultivar, which was used to derive improved cultivar such as Zatib (see Table 3.5 for cultivars's genetic coefficients). A brief description of the cultivars' genetic coefficients can be found in Appendix A.2.

3.2.2 *Simulated Results*

3.2.2.1 Rice

The simulated yields in both 2050 and 2080 were done under two climate scenarios: 'left low emissions' or RCP 2.6 and 'left high emissions' or RCP 8.5. A fertilizer amount of 100 kg/ha urea-nitrogen has been applied in all simulations. Table 3.6 summarizes the predicted yields of rice in 2050 and 2080 based on the rice cultivar that we used in Chap. 2 (IR 64) and the two additional improved cultivars (NE4 and DK3), from which we observe higher yields in both years and under both climate scenarios. Several other observations can be made: (i) higher yields are found in the low emissions climate scenario than in the high emissions climate scenario, (ii) NE4 cultivar seems to be more resilient to climate change since the highest yields are obtained with that cultivar, and (iii) in 2050 rice yields are higher in Tambacounda for NE4 and DK3, while in 2080 yields obtained in Guiragor are the highest.

As done in Chap. 2, the simulated production is the product of the simulated yields and the planting area change scenarios (see Table 2.11). As a result, there is a significant improvement in the predicted rice production with the two alternative cultivars as shown in Fig. 3.2, regardless of the climate scenario. In 2050, the highest rice production reaches 1.22B kg when using DK3 rice cultivar when there is an

Table 3.6 Rice simulated yields under IR64, NE4 and DK3 cultivars

	2050					
	IR 64		NE4		DK3	
Region	RCP 2.6	RCP 8.5	RCP 2.6	RCP 8.5	RCP 2.6	RCP 8.5
Yang Yang	190	361	678	962	697	409
Tambacounda	1593	1313	4085	3402	4041	3869
Guiragor	4266	3856	4082	3998	3431	2766
Senegal (Aggregate level)	2016	1843	2948	2787	2723	2348
	2080					
	IR 64		NE4		DK3	
Region	RCP 2.6	RCP 8.5	RCP 2.6	RCP 8.5	RCP 2.6	RCP 8.5
Yang Yang	656	0	1112	211	250	57
Tambacounda	1555	369	3820	1526	3507	1270
Guiragor	4266	2231	4084	3591	3431	2199
Senegal (Aggregate level)	2159	867	3005	1776	2396	1175

Source Author's computations

increase of 10% in planting area and RCP 2.6 climate scenario. While the lowest production is 676M kg if the rice cultivar used in Chap. 2 (IR 64) was maintained combined with a 10% decrease in planting area and RCP 8.5. In 2080, DK3 rice cultivar outperformed the others with a maximum production of 1.48B kg under RCP 2.6 and a 10% increase in planting area. The lowest production is still obtained with IR64 cultivar under RCP 8.5 and 10% decrease in planting area (286M kg). Appendix D tabulates the simulated production results.

3.2.2.2 Maize

Simulated maize yields at study sites and at the national level are summarized in Table 3.7. Similar to the rice simulations above, simulated yields in both 2050 and 2080 were done under two climate scenarios: "left low emissions" or RCP 2.6 and "left high emissions" or RCP 8.5. A fertilizer amount of 100 kg/ha urea-nitrogen has been applied in all simulations. While "left Massongo" is the maize cultivar that was used in Chap. 2, TZE COMP4 C2 and TZB-SR are the improved cultivars as a result of technological changes. While the highest yield in 2050 is obtained with Massongo (4,182 kg/ha) under low emissions climate change scenario, the highest yield in 2080 is 4,535 kg/ha and is obtained by using one of the improved cultivars TZB-SR under RCP 2.6. This might highlight the resilience of Massongo cultivar, which seems to resist climate shocks and dry conditions. At the regional level, the highest yields are found in the central sub-humid region (Tambacounda), closely followed by the south

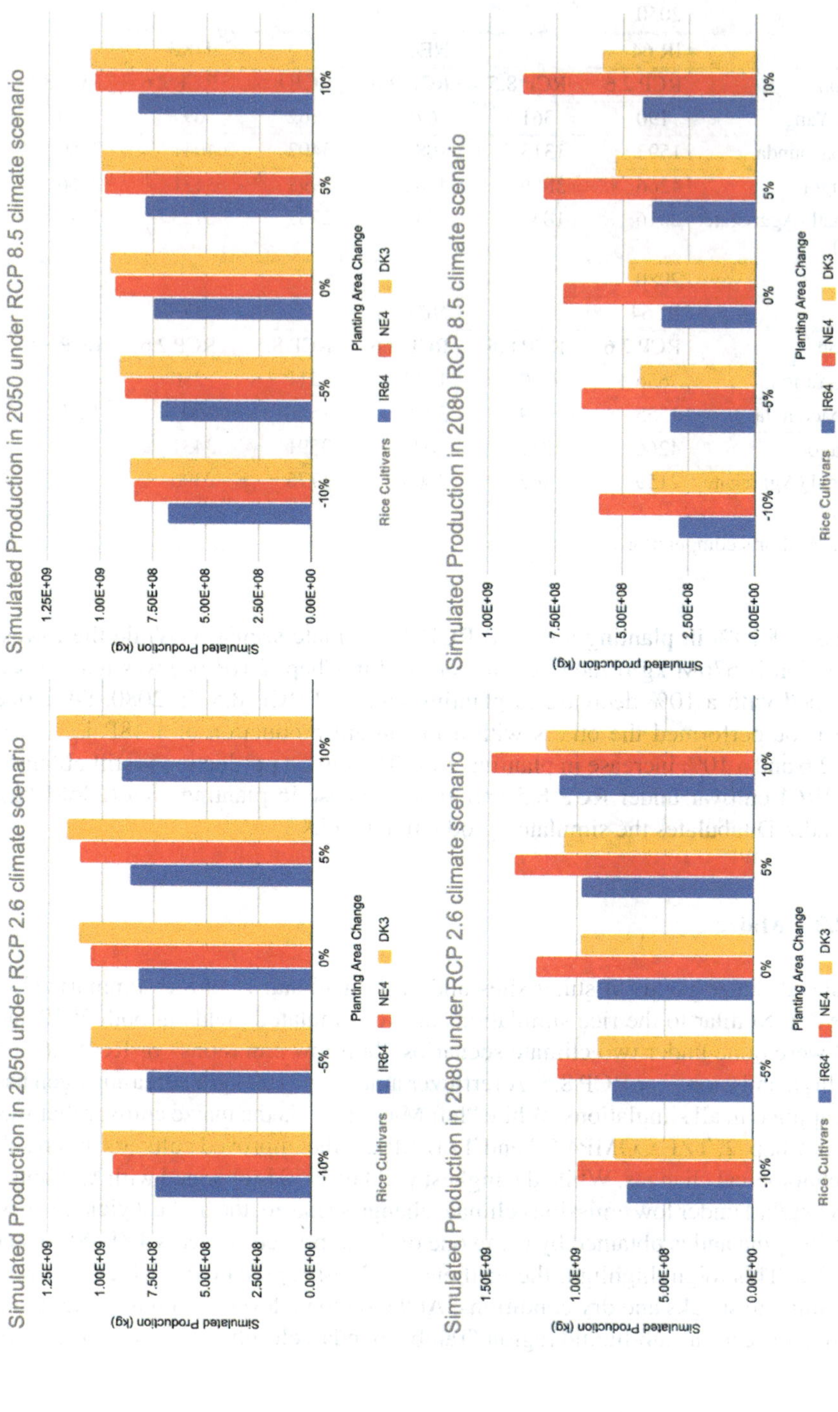

Fig. 3.2 Rice simulated production under IR64, NE4 and DK3 Cultivars. *Source* Author's computations

Table 3.7 Maize simulated yields under Masongo, TZE COMP 4C2, and TZB-SR cultivars

	2050					
	Massongo		TZE COMP 4C2		TZB-SR	
Region	RCP 2.6	RCP 8.5	RCP 2.6	RCP 8.5	RCP 2.6	RCP 8.5
Yang Yang	2271	2327	2115	1970	2396	2560
Tambacounda	5861	5157	5035	4649	5535	5354
Guiragor	4413	4006	4309	4092	4535	4161
Senegal (Aggregate level)	4182	3830	3820	3570	4155	4025
	2080					
	Massongo		TZE COMP 4C2		TZB-SR	
Region	RCP 2.6	RCP 8.5	RCP 2.6	RCP 8.5	RCP 2.6	RCP 8.5
Yang Yang	2372	892	2261	1202	2607	1076
Tambacounda	5657	2840	5078	3122	5314	3601
Guiragor	4413	3088	4309	3327	4535	3608
Senegal (Aggregate level)	4147	2273	3883	2550	4152	2762

Source Author's computations

humid region (Guiragor), and the north arid (Yang Yang) where very low yields are recorded.

As a result of the simulated yields variations and the different planting area change scenarios (Table 2.11), the highest predicted maize production is 1.32B kg in 2050, when Massongo cultivar is used under "low emissions" climate scenario and a 10% increase in planting area. This implies that both improved cultivars (TZB-SR and TZE COMP 4C2) are not as efficient as Massongo in 2050. One can see from Fig. 3.3 and Table D.2 (in Appendix D) that the lowest maize production in that same year is roughly 922M kg with TZE COMP 4C2 under RCP 8.5 and a 10% decrease in planting area. However, in 2080, simulated maize production reaches 1.44B kg with TZB-SR cultivar under "low emissions" climate scenario and a projected 10% increase in planting area. Indeed, simulated maize production in 2080 has the lowest values with Massongo cultivar under both "low emissions" and "high emissions" climate scenarios.

3.2.2.3 Sorghum

Based on the reported yield simulation results in 2050, both improved sorghum cultivars (improved CSM335 and Fadda) have higher yields than the cultivar used in the previous chapter under both RCP 2.6 and RCP 8.5 climate change scenarios. Sorghum seems to have higher productivity in the south humid region (Guiragor) than in the central sub-humid region (Tambacounda) as opposed to maize and rice. Similar patterns were found in 2080, where improved cultivars have higher yields

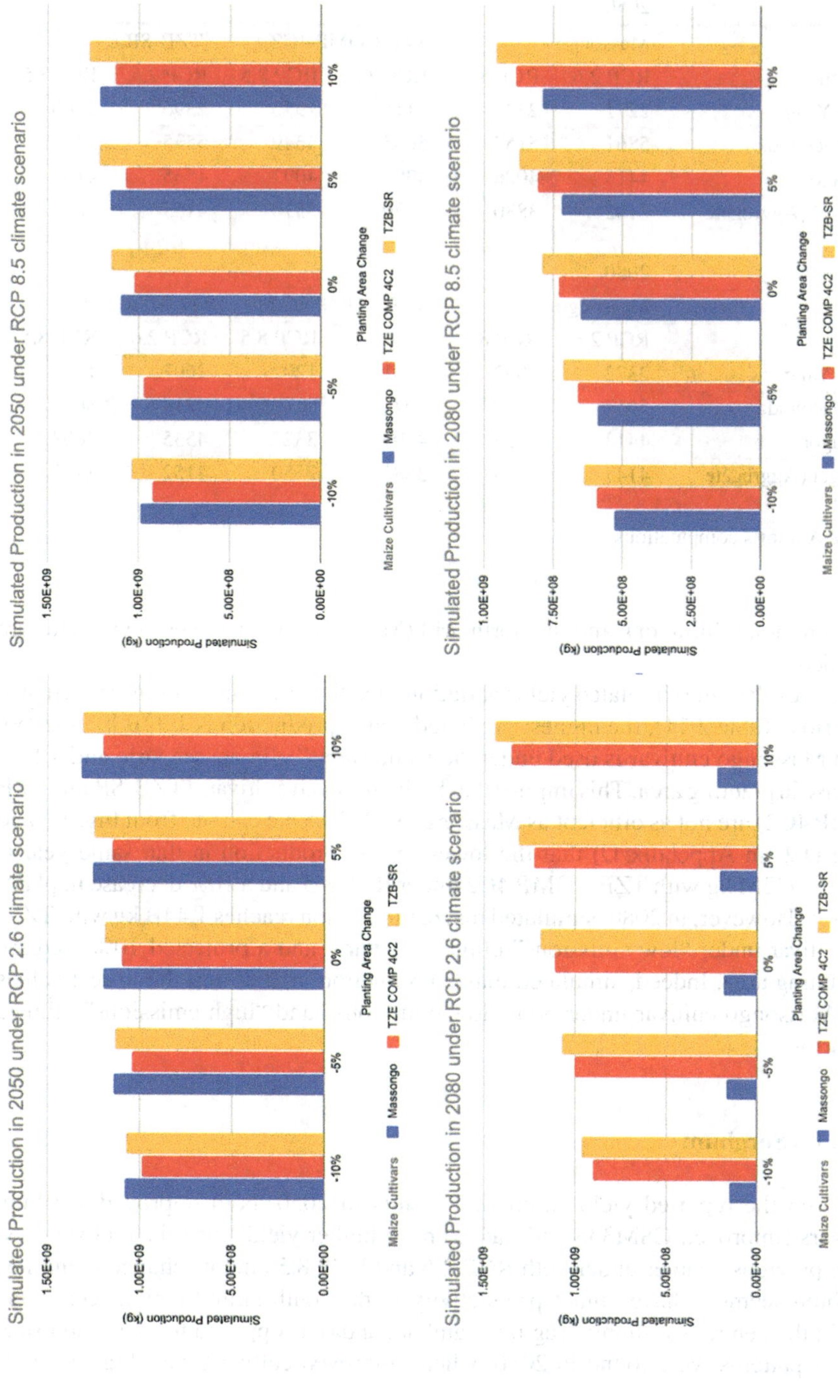

Fig. 3.3 Maize simulated production under Masongo, TZE COMP 4C2, and TZB-SR cultivar. *Source* Author's computations

Table 3.8 Sorghum simulated yields under CSM335, improved CSM335, and Fadda Cultivars

	2050					
	CSM335		Improved CSM335		Fadda	
Region	RCP 2.6	RCP 8.5	RCP 2.6	RCP 8.5	RCP 2.6	RCP 8.5
Yang Yang	1399	1482	1661	1402	1495	1359
Tambacounda	2856	2895	2952	2947	2748	2776
Guiragor	3172	2922	3224	3230	3313	3128
Senegal (Aggregate level)	2476	2433	2612	2526	2519	2421
	2080					
	CSM335		Improved CSM335		Fadda	
Region	RCP 2.6	RCP 8.5	RCP 2.6	RCP 8.5	RCP 2.6	RCP 8.5
Yang Yang	1452	1080	1657	799	1499	783
Tambacounda	2623	1815	2946	2198	2617	2007
Guiragor	3164	2709	3165	2927	3279	2777
Senegal (Aggregate level)	2413	1868	2589	1975	2465	1856

Source Author's computations

than the CSM335 cultivar used in Chap. 2. Also, Guiragor is the region where the highest sorghum yields are predicted. The improved CSM335 cultivar is the most high-achieving yield cultivar, which seems to be more drought tolerant in both years (see Table 3.8).

As done with rice and maize, the simulated production of sorghum is the product of the simulated yields and the different planting area change scenarios (Table 2.11). In both 2050 and 2080, the highest production of sorghum is achieved when the improved CSM335 cultivar is used. In 2050, the highest production of sorghum is estimated at around 800M kg under "low emissions" climate scenario and 10% increase in planting area. The lowest production in that same year is roughly 606M kg with the CSM335 cultivar used in Chap. 2 under "low emissions' climate scenario given a 10% decrease in planting area. In 2080, the same improved CSM335 cultivar is expected to achieve a production of around 871M kg under "low emissions' climate scenario given a 10% decrease in planting area as shown in Fig. 3.3 and Table D.2 (in Appendix D).

3.2.2.4 Pearl Millet

As highlighted in Table 3.8, yield simulation results in 2050 of both improved pearl millet cultivars (Zatib and Heini Kirey) have higher yields than the CIVT cultivar used in the previous chapter under both RCP 2.6 and RCP 8.5 climate change scenarios. Similar to sorghum and as opposed to rice and maize, pearl millet seems to have higher productivity in the south humid region (Guiragor) than in the cen-

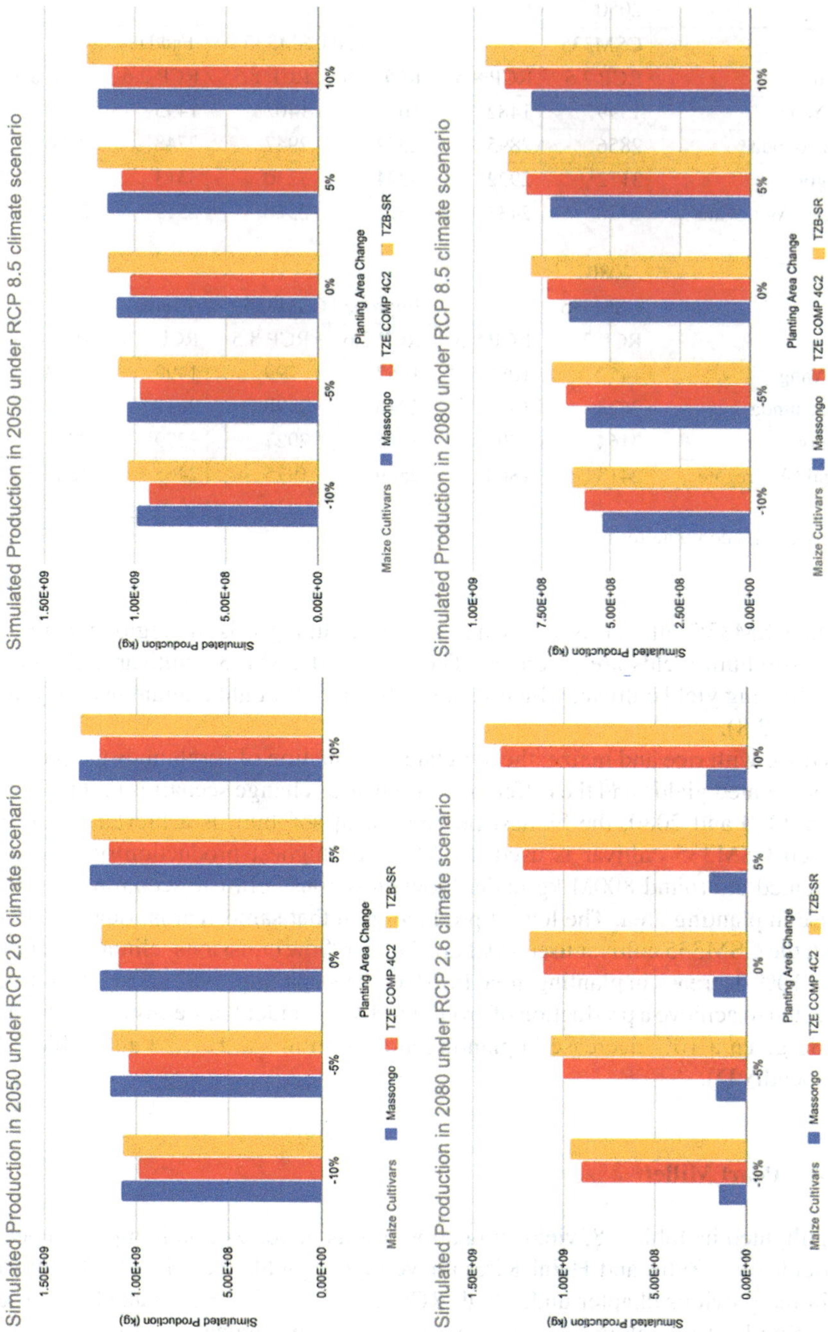

Fig. 3.4 Sorghum simulated production under CSM335, improved CSM335, and Fadda cultivars. *Source* Author's computations

tral sub-humid region (Tambacounda). Likewise patterns were found in 2080, where improved cultivars have higher yields than the CIVT cultivar used in Chap. 2. Also, Guiragor is the region where the highest pearl millet yields are predicted. Under the "low emissions" climate scenario in 2050 and 2080, Heini Kirey had the higher yields followed by Zatib, then CIVT. Whereas under the "high emissions" climate scenario, Zatib cultivar had the highest yields in both 2050 and 2080.

As computed with the other crops, the simulated production of pearl millet is the product of the simulated yields and the different planting area change scenarios. In both 2050 and 2080, the highest production of pearl millet is achieved when the improved cultivars are used. In 2050, the highest production of pearl millet is approximately at around 2B kg under the "high emissions" climate scenario and 10% increase in planting area when Zatib is used. The lowest production in that same year is roughly 1.25B kg with the CIVT cultivar used in Chap. 2 under "low emissions' climate scenario given a 10% decrease in planting area. In 2080, the Heini Kirey cultivar is expected to achieve a production of around 2.13B kg under "low emissions' climate scenario given a 10% decrease in planting area as shown in Fig. 3.4 and Table D.3 (in Appendix D) (Table 3.9 and Fig. 3.5).

3.2.2.5 Aggregate Results

In order to better assess the impact of new technologies (with improved crop cultivars) in the aggregate cereal production, we are going to compare the maximum and the minimum aggregate cereal production when the improved cultivars are used and when

Table 3.9 Pearl Millet simulated yields under CIVT, Zatib, and Heini Kirey Cultivars

	2050					
	CIVT		Zatib		Heini Kirey	
Region	RCP 2.6	RCP 8.5	RCP 2.6	RCP 8.5	RCP 2.6	RCP 8.5
Yang Yang	519	632	750	824	714	593
Tambacounda	1454	2131	1963	2484	1824	1642
Guiragor	2100	1914	2396	2121	2791	2623
Senegal (Aggregate level)	1358	1559	1703	1810	1776	1619
	2080					
	CIVT		Zatib		Heini Kirey	
Region	RCP 2.6	RCP 8.5	RCP 2.6	RCP 8.5	RCP 2.6	RCP 8.5
Yang Yang	507	233	702	248	714	273
Tambacounda	1152	629	1526	1127	1694	789
Guiragor	2100	1515	2396	1782	2747	1768
Senegal (Aggregate level)	1253	792	1541	1052	1718	943

Source Author's computations

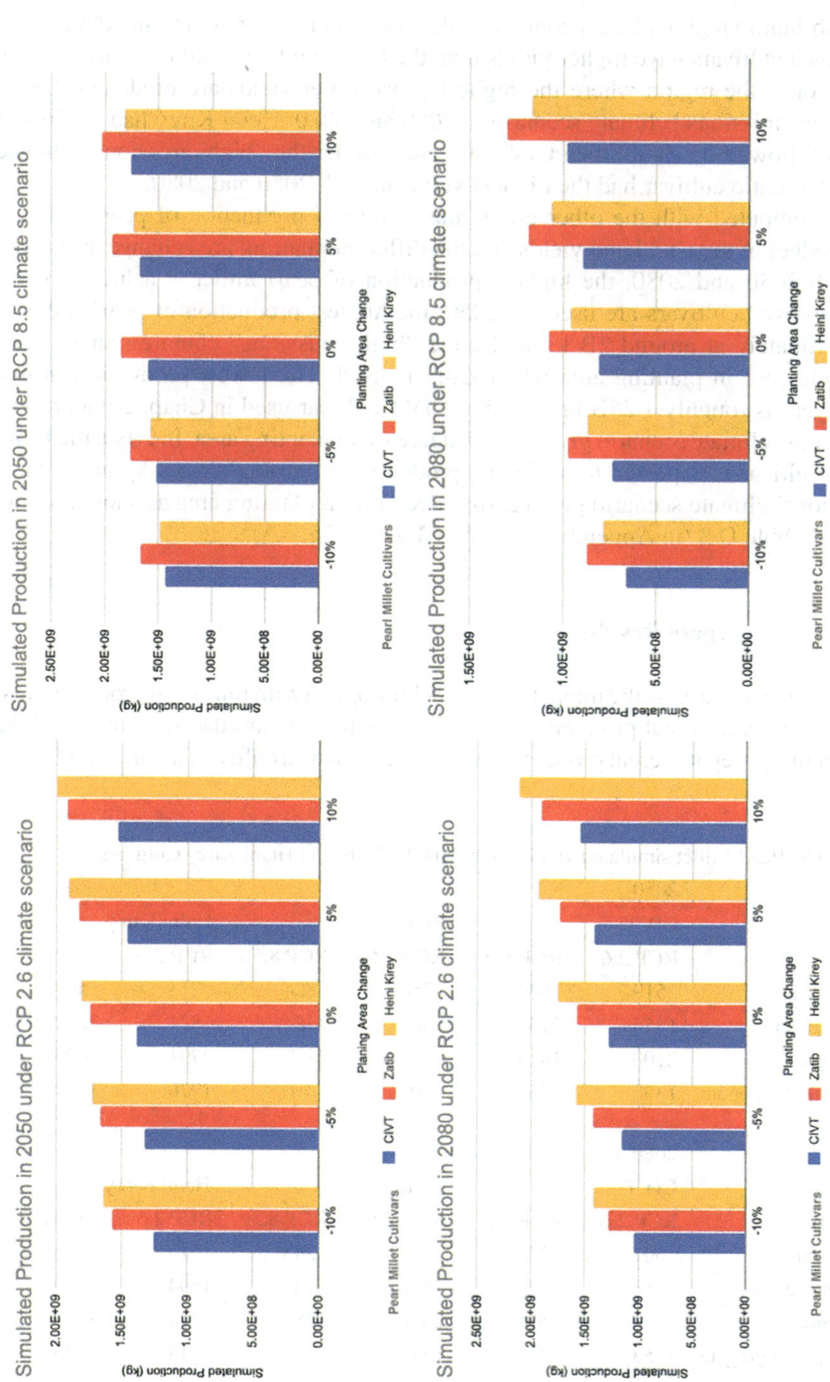

Fig. 3.5 Pearl Millet simulated production under CIVT, Zatib, and Heini Kirey Cultivars. *Source* Author's computations

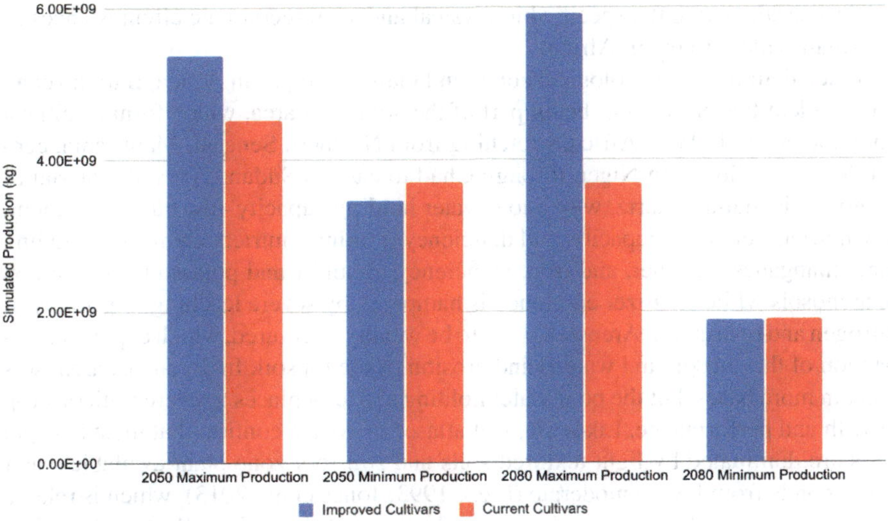

Fig. 3.6 Simulated maximum and minimum Cereal production with improved cultivars (in blue) and current cultivars (in red). *Source* Author's computations

the cultivars from Chap. 2 are maintained. Our main takeaway from the obtained results (summarized in Fig. 3.6 and Table D.5 in Appendix D) is that with the use of more updated or well adapted cultivars—such as TZE COMP 4C2 (for maize), improved CSM335 (for sorghum), Zatib and Heini Kirey (for pearl millet), and NE4 and DK3 (for rice)—the maximum cereal production in 2050 could be increased by 857 million kg and the minimum production could also be decreased by 256 million kg. Likewise in 2080, the maximum cereal production could be increased by 2.24 billion kg and the minimum production could be decreased by 17.2 million kg. These results confirm that new technologies via new varieties of crop cultivars can help farmers adapt to the abiotic stresses of climate change and mitigate climate shocks impacts by improving the drought and heat tolerance of the studied crops.

3.3 Changing Quality of Soil

3.3.1 Senegalese Current Soil Properties

"A nation that destroys its soils destroys itself."—Franklin Roosevelt. Sixty five percent of the agricultural land in Africa is degraded and soil fertility depletion, a manifestation of soil degradation, is currently a serious threat to food security among small-holder farmers (Bationo et al., 2003). There is, therefore, a need for agricultural intensification through efficient use of soil nutrient and water resources. Technologies need to be adapted to the specific biophysical and socio-economic circumstances of the small scale farmers in Africa.

When defining agro ecological zones and main soil types in Africa, Bationo et al. (2003) identified Senegal as being part of the arenosols area, which form an almost continuous belt in West Africa, stretching from Northern Senegal, Mauritania, central Mali, and Southern Niger, through Chad to eastern Sudan. The soil material of arenosols is mainly quartz, with a low water holding capacity, low nutrient content, low nutrient retention capacity, and deficiency of minor nutrient elements including zinc, manganese, copper, and iron. Deficiency of sulfur and potassium is common in arenosols while fertilizer efficiency is hampered by severe leaching, especially of nitrogen and potassium. Arenosols tend to be weakly structured, which explains compaction of the subsoil and water/wind erosion of the topsoil. In dry areas, arenosols contain more bases, but the poor water holding capacity places a severe limit on crop growth and performance. Likewise, soil atlases of Africa confirm that most Senegal soils are dominated by light textured soils and soil root zone plant available water corresponds from low to moderate (FAO, 1993; Jones et al., 2013), which is related to the sand–sandy loam soils. Consequently, most of the soils in Senegal have low water holding capacity.

3.3.2 *Predicted Soil Profile*

In our simulations, pests, diseases, and weeds are assumed to be managed. The simulation study was conducted on a dominant soil type (sand) in Senegal, which may not represent all specific locations. We know from Bationo et al. (2003), Jones et al. (2013), and FAO that crops grown on sandy soils are sensitive to long-dry spells as the soils have low water holding capacity. Therefore, we are going to consider sand soil as the worst-case scenario, which could help explore the potential impacts of climate change and adaptation management strategies. In this worst-case scenario, crops are exposed to grow in soils that retain less water (sand). Crops grown on low water holding capacity soils suffer from limited water (drought) stress which could have several negative effects on crops such as nutrient uptake, photosynthesis and translocation, hindering growth and development and reproduction function. Some studies also showed that considering the worst scenario is important for developing adaptation strategy (Araya et al., 2021). In this section, our simulations of future climate change do not account for temperature extremes that will have greater negative impacts if they occur during sensitive stages of crop development.

Niger sandy soil is the West African soil that corresponds to the sandy soil texture adopted from the DSSAT soil database with profile name: IBML910001. The soil input information includes: (i) general site, (ii) soil surface information, and (ii) soil profile characteristics. The general site information is about latitude, longitude, and elevation. The soil surface information requires soil taxonomy (if available), soil slope, soil color, and stones (%). Lastly, the soil profile characteristics contains soil texture (% sand, silt, and clay and stones), soil organic carbon, bulk density,

*IBML910001	IBSNAT	-99	210	Niger Sandy Soil

@SITE	COUNTRY	LAT	LONG	SCS FAMILY
-99	Niger	-99	-99	NIGER SANDY SOIL

@ SCOM	SALB	SLU1	SLDR	SLRO	SLNF	SLPF	SMHB	SMPX	SMKE
-99	.14	10	.8	76	.35	.7	IB001	IB001	IB001

@ SLB	SLMH	SLLL	SDUL	SSAT	SRGF	SSKS	SBDM	SLOC	SLCL	SLSI	SLCF	SLNI	SLHW	SLHB	SCEC	SADC
5	-99	.031	.097	.32	1	-99	1.61	.5	-99	-99	-99	-99	5.5	-99	-99	-99
15	-99	.031	.097	.32	1	-99	1.61	.5	-99	-99	-99	-99	5.5	-99	-99	-99
30	-99	.037	.119	.331	.8	-99	1.66	.5	-99	-99	-99	-99	5.5	-99	-99	-99
45	-99	.037	.117	.332	.6	-99	1.54	.5	-99	-99	-99	-99	5.5	-99	-99	-99
60	-99	.037	.117	.332	.4	-99	1.54	.5	-99	-99	-99	-99	5.5	-99	-99	-99
90	-99	.04	.126	.331	.01	-99	1.57	.5	-99	-99	-99	-99	5.5	-99	-99	-99
120	-99	.037	.117	.331	.01	-99	1.55	.5	-99	-99	-99	-99	5.5	-99	-99	-99
150	-99	.037	.119	.33	.01	-99	1.64	.5	-99	-99	-99	-99	5.5	-99	-99	-99
180	-99	.037	.119	.331	.01	-99	1.58	.5	-99	-99	-99	-99	5.5	-99	-99	-99
210	-99	.034	.108	.324	.01	-99	1.6	.5	-99	-99	-99	-99	5.5	-99	-99	-99

Fig. 3.7 Listing of sample soil profile of Niger sandy soil with required data in red. *Source* Decision Support System for Agrotechnology Transfer (DSSAT) version 4.8 data

Lower Limit of plant extractable soil water (LL), and Drained Upper Limit (DUL) or field capacity (see Fig. 3.7 for Niger sandy soil profile). Complete definitions of soil parameters in DSSAT soil file can be found in Table 2.4.

3.3.3 Simulated Results

As expected, simulated yields in the worst case scenario record the lowest values, especially in the "high emissions" climate change scenario. However, an interesting finding is that in the north-arid region (Yang Yang), as shown in Table 3.10, there is a substantial yield increase, both in 2050 and 2080, when the Niger sandy soil is applied, regardless of the climate change scenario (RCP 2.6 and RCP 8.5). This might imply that the dry locations in the northern part of the country would be more suitable for cereal production than the other parts of the country.

One reason for this might be that the central sub-humid and south-humid parts of the country are highly vulnerable to the reduced available soil water because of reduced precipitation during the growing season from shortening of growing season and increased daily evapotranspiration. Some studies showed that increased CO_2 and rise in temperatures could limit yield (Jayawardena et al., 2017). Another plausible reason for these changes is attributed to the impact of projected warming which could alter biogeochemical processes in plants and soil, microbial respiration, and hence indirect negative impact on crop yields (Melillo et al., 2011). As a result, simulated cereal production drastically decreases at the national level in both 2050 and 2080, with a further decrease in the "high emissions" climate scenario or RCP 8.5 (as indicated in Fig. 3.8 and in Appendix D).

Table 3.10 Aggregate simulated cereal yields based on soil profile

	2050				2080			
	Baseline yield (Results from Chap. 2)		Niger Sandy Soil profile yield (Worst-case scenario)		Baseline yield (Results from Chap. 2)		Niger sandy soil profile yield (Worst-case scenario)	
Region	RCP 2.6	RCP 8.5	RCP 2.6	RCP 8.5	RCP 2.6	RCP 8.5	RCP 2.6	RCP 8.5
Rice								
Yang Yang	190	361	480	746	656	0	958	0
Tambacounda	1593	1313	772	800	1555	369	713	392
Guiragor	4266	3856	479	456	4266	2231	479	421
Senegal (Aggregate level)	2016	1843	577	667	2159	867	717	271
Maize								
Yang Yang	2271	2327	2995	2920	2372	892	2788	1413
Tambacounda	5861	5157	2725	2632	5657	2840	2672	1548
Guiragor	4413	4006	3317	2717	4413	3088	3317	2640
Senegal (Aggregate level)	4182	3830	3012	2756	637	2273	2926	1867
Sorghum								
Yang Yang	1399	1482	1970	2065	1452	1080	1975	1422
Tambacounda	2856	2895	1399	1464	2623	1815	1272	1195
Guiragor	3172	2922	2219	2164	3164	2709	2223	2116
Senegal (Aggregate level)	2476	2433	1863	1898	2413	1868	1823	1578
Pearl millet								
Yang Yang	519	632	872	1046	507	233	825	386
Tambacounda	1454	2131	257	842	1152	629	256	268
Guiragor	2100	1914	771	886	2100	1515	771	790
Senegal (Aggregate level)	1358	1559	633	925	1253	792	617	481

Source Author's computations

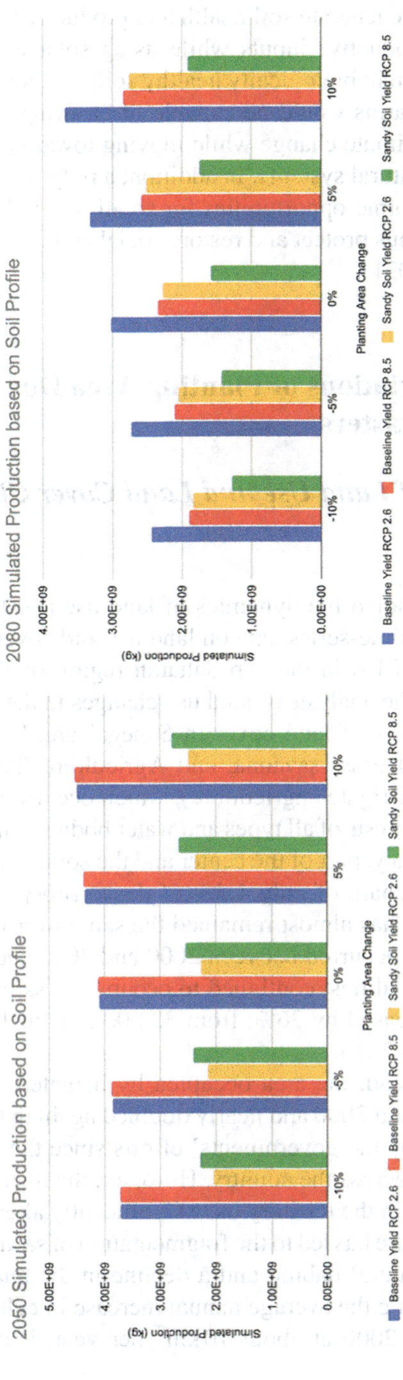

Fig. 3.8 Aggregate simulated cereal production based on soil profile. *Source* Author's computations

As a solution to improve soil profile, diversified, polycultural farms have the capacity to restore and regenerate soil health and productivity by transforming agricultural wastes into productive inputs while using solar energy to sequester and store atmospheric carbon in biologically healthy soils. A global agriculture made up of small, polycultural farms would be capable of ensuring global food security in the face of inevitable climate change while moving toward carbon-neutral, or even carbon-negative, agricultural systems. In addition, a polyfunctional agrifood system would create new economic opportunities for small, individually owned and operated family farms and thus protect and restore social and economic integrity in rural communities (Batini, 2021).

3.4 Predicted Variations of Planting Area Due to Longer Persistent Disasters

3.4.1 Dynamics of Land Use and Land Cover Changes Over the Past

Tappan et al. (2016) studied the dynamics of land use and land cover in Senegal. Their project produced time-series data on land use and cover in three distinct periods (1975, 2000, and 2013) in the sub-Saharan region of West Africa, including Senegal, which allows the analysis of land use changes in the region. Based on their findings, the main classes of land cover in Senegal are, by order of importance, Savanna, Sahelian short grass savanna, and Agriculture (including agriculture in shallow, recession, and irrigated agriculture), which occupy 35, 26, and 22% of the territory, respectively. Forests of all types and water bodies, which are mainly located in the Senegal River Valley, part of the center and the south, occupy a negligible part of the territory, with a share of only 1.3% of the territory for each. Furthermore, this land cover partition has almost remained the same over the period 1975–2013. The main changes have occurred between 2000 and 2013 and touched mainly total agricultural land, which almost continued to occupy the same space between 1975 and 2000 and then increased by 26%, from 32,600 km^2 in 1975 to 32,900 km^2 in 2000 and 41,000 km^2 in 2013.

Within agricultural land, the area occupied by irrigated agriculture has almost doubled between 1975 and 2000 and nearly doubled again in the period 2000–2013. This can be explained by the governments' efforts since the 1970s-era droughts to increase irrigated areas across the country. However, the distribution and expansion of agricultural land within the territory have significantly altered the landscape. The development of agriculture has led to the fragmentation of savannah and open forests, resulting in a loss of natural habitat and a decline in the quality of the remaining natural ecosystems. While the average annual increase in cultivated area was rather low between 1975 and 2000 at about 10 km^2 per year, it increased dramatically

Table 3.11 Main causes of land degradation in senegal

Geographical location	Causes of land degradation
North Arid (Yang Yang)	1. Tree cuts for charcoal, bush fires, and crop area expansion 2. Abusive exploitation of firewood, clearing for agricultural purposes 3. Drop in rainfall
Central Sub-Humid (Tambacounda)	1. Widespread decrease in rainfall amounts 2. Abusive land clearing for agricultural purposes
South Humid (Guiragor)	1. Decreased rainfall 2. Farming activities 3. Land clearing, logging, bush fires 4. Exploitation of rice-growing areas

Source "Environnement et Gestion Des Ressources Naturelles" by ANAT (2016)

between 2000 and 2013, to an average of 630 km^2 per year. However, this trend masks disparities within the Senegalese territory (CILSS, 2016).

In terms of space occupied, agricultural land increased by 4% between 1975 and 2013, with more changes occurring between 2000 and 2013 as previously stated. The space occupied by Savana decreased by around 3% between 2000 and 2013. As for Savana short grass, changes are negligible. The occupation of forests, on the contrary, have been declining over time, decreasing from 1.66% to 1.43 and 1.28% in 2000 and 2013, respectively.

About the dynamics of land degradation, studies carried out by the CILSS in November 2010 indicate that, out of the 3,805,000 ha of arable land available in the country, 2,400,000 ha, or 63%, are severely degraded (IED, 2015). Within the same range, (i) IRD (2016) shows that soil degradation affects 2/3 or 67% of the country's arable land, generating a loss equivalent to 1% of the GDP, and (ii) the Senegal Country Environmental Assessment (IRD, 2016) reports that nearly 65% of total agricultural land is degraded. A CILSS study (CILSS, 2016) shows land degradation in the Ferlo region (central Senegal) between 1994 and 2011 as well as a decline in vegetation cover and biodiversity in central-eastern Senegal between 1984 and 2013. Moreover, the LDN report (ANAT 2016; MEDD, 2018a; WorldBank 2008) and show that at the biophysical level, the land area of Senegal measures 20,179,118 ha, of which the Project 'Assessment of Land Degradation in Drylands' estimates the level of land degradation at 34%, which represents a degraded area of 6,860,900 ha (Table 3.11).

The intensification and expansion of land degradation can severely affect labor productivity and jeopardize agricultural livelihoods in the country (MEDD, 2018a). Indeed, land degradation can have serious impacts on livelihoods by limiting the availability of vital ecosystem services (deterioration of food and water availability, soil fertility), increasing the risk of poverty and forcing people to migrate. Between 2000 and 2010, the number of people living in remote agricultural areas undergoing degradation and with limited market access increased by 37%, reaching 257,000 people (MEDD, 2018a).

3.4.2 Agrimonde-Terra Scenarios of Land Use

Issues of land use and food and nutritional security are marked by a high degree of uncertainty, where it is difficult to consider all the possible consequences of an action because of the interdependence of factors and a multiplicity of values and actors that are present in the definition of the problem and its solution. Answering one aspect of the problem may lead to or generate other problems because the dynamics are interlocked and the system is constantly changing. So the future relationship between land use and food security cannot be reduced to a single issue, for example population growth, but it is necessary to take into account changes in agricultural, environmental, climate, and food systems etc.

Therefore, in order to simulate different scenarios of land usage, we will work with the Agrimonde-Terra's foresight approach to scenario construction developed by Mora and DeLattre (2018), which offers both a reformulation of the problem of the relationship between land use, the impact of natural disasters, and food security by 2050, identifies the issues to be addressed. The construction of the scenarios took place in four steps: (1) the definition of the 'land use and food security' system; (2) the construction of the alternative hypotheses of changes to 2050 for each driver of the system; (3) the development of land use and food security scenarios by combining hypotheses and the construction of narratives describing the scenarios; and (4) the implementation of simulations illustrating the quantified impact of the qualitative assumptions on land use and food security as a result of changes in environmental, climate. As highlighted in Table 3.12, Mora and DeLattre (2018) identified the following scenarios and their predicted impact on planting area in 2050: (i) metropolization, regionalization, healthy, and communities scenarios.

3.4.3 Simulated Results

The simulated cereal production for each Agrimonde-Terra's land usage scenario is computed by taking the product of the simulated yield with fertilizer results reported in Chap. 2 (Table 2.10) and the resulting area for each crop (which is obtained by applying the percentage area change to the 2020 baseline area of each crop, as indicated in Appendix D in Table D.7).

Each of the Agrimonde-Terra scenarios has its own set of assumptions. In the Communities scenario, it is assumed that in sub-Saharan Africa (SSA), some farmers will organize themselves, develop agro ecological farms, and manage land so as to provide foodstuffs, energy, and environmental services. Other farmers will fail to organize themselves and become more marginalized. There will, therefore, be inequity at continental and national levels and a struggle for resources. As a result, the simulated cereal production varies between 4.8B and 5.6B kg in 2050, while in 2080 it is between 3.25B to 5.6B kg (see Fig. 3.9 and Tables D.8 and D.9).

Table 3.12 Agrimonde-Terra's scenario descriptions and land use change in 2050

Scenario	Variant (assumptions used)	Arable and permanent crops area change (%)
Metropolization	(a) With ultra-processed products: transition to diets based on ultra-processed products	16
	(b) With animal products: transition to diets based on animal products	40
Regionalization	(a) With agricultural technology A: sustainable intensification for cropping systems + Conventional intensification with local resources for livestock systems	4.5
	(b) With agricultural technology B: Agroecology for cropping systems + Agroecological livestock	11
Healthy	(a) With agricultural technology C: sustainable intensification for cropping systems +Agroecological livestock	–4
	(b) With agricultural technology D: Agroecology for cropping systems + Agro Ecological livestock	3
Communities	(a) With agroecology: agroecology for cropping systems + Agro Ecological livestock	18
	(b) With collapse: collapse of crop production + Backyard livestock	36

Source "Land use and food security in 2050: a narrow road" by Mora and DeLattre (2018)

If the Metropolization scenario develops the cropland area increases at the world level in both variants. This explains the high production cereal values which fluctuate between 4.8B and 5.8B kg in 2050 and between 3B and 6B kg in 2080. Once again, the 'left with ultra-processed products' variant requires less cropland (+16%) than the 'left with animal products' variant (+40%). In both variants, it is important to note that regions in West Africa are constrained by their maximum cultivable area (see Fig. 3.9 and Tables D.8 and D.9).

With the Regionalization scenario, Africans will become the primary beneficiaries of their production. However, given the demand, the reduction of imports, especially of animal product imports, is likely to lead to considerable deforestation. This leads to a fall of the simulated cereal production in both 2050 (between 4.3B and 4.6B kg) and 2080(between 2.5B and 3.8B kg) compared with the metropolization and communities scenarios (see Fig. 3.9 and Tables D.8 and D.9).

Finally, the Healthy scenario requires major and difficult political changes as well as changes in diets in SSA; it leads to an expansion of agricultural areas, but also to carbon sequestration in soils; it can contribute to employment in the agrifood sector; therefore, it will be globally beneficial for land use and food security but not without challenges especially in West Africa where, in 2050, arable land and permanent crops will be close to the maximum cultivable area. Hence as shown in Fig. 3.9 and Tables D.8 and D.9, the 2050 simulated cereal production is at around 4B kg and is expected to further decrease to 3B kg.

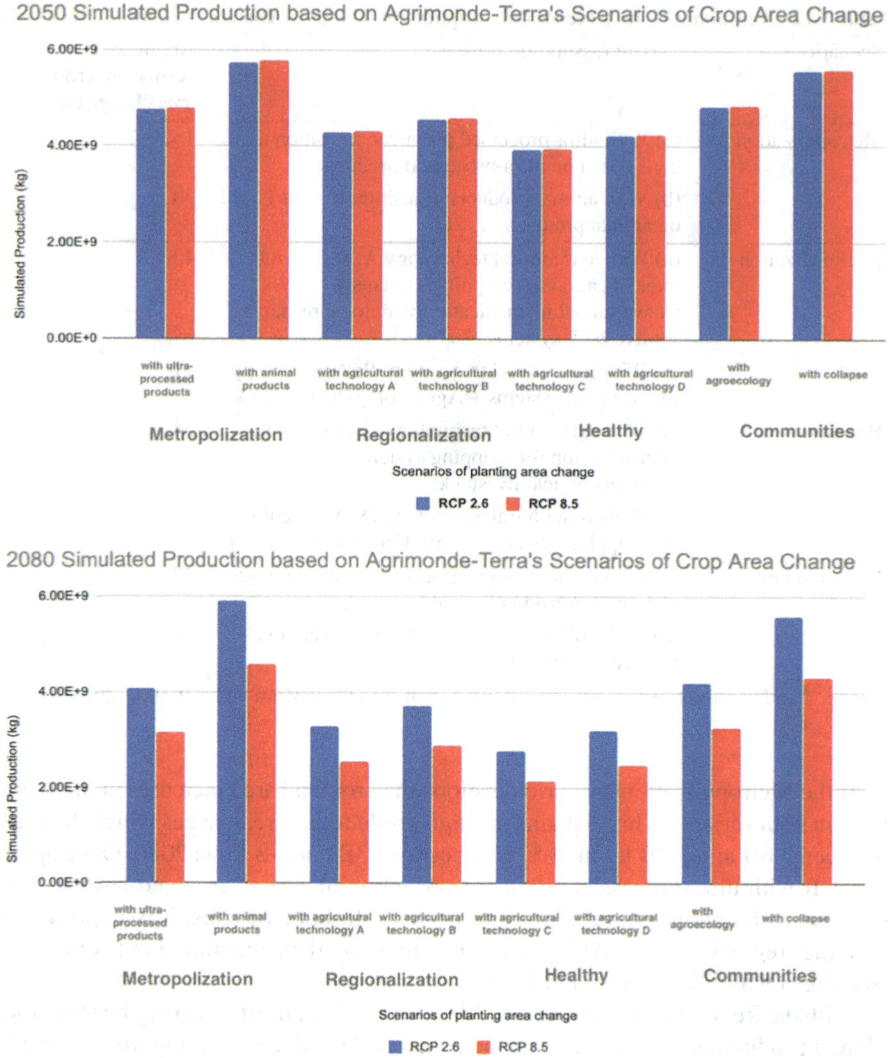

Fig. 3.9 Agrimonde-Terra simulated Cereal production in 2050 and 2080. *Source* Author's computations

It is worth noting that the reason we chose the Agrimonde-Terra land usage scenarios is that the incidences of drought have declined significantly within the most recent decade. However, it remains unclear whether this change is temporary or rather the result of a shift in long-term weather patterns. Although an observed wetting over roughly the same period has led to recurrent flooding in many parts of the country, affecting urban and coastal areas, impacts on crop and livestock production have been limited at the aggregate level.

The simulated cereal production results obtained in Chap. 2 and this chapter highlights that like most of the countries in the Sahel region, Senegal's agricultural sector is highly vulnerable to the effects of climate change. The country's climate is already characterized by high temperatures and highly variable annual precipitation, factors that often negatively affect the productivity of heat-sensitive crops. More than 95% of the total cropped area depends on rainfed systems, and most farmers and herders practice traditional forms of agriculture. Because of these factors and the importance of the sector in Senegal's national economy, climate change impacts on crop yields and land suitability will have far-reaching effects.

Uncertainties surrounding the degree of temperature increases and the direction of change in precipitation levels make it difficult to determine the precise impact of future climate change on crop yields. Predictions of crop yields vary depending on the underlying assumptions, and most models, including the one we used, do not account for the way in which the changing incidence of pests and diseases will affect farm productivity. Despite those uncertainties, we found that the overall impacts of climate change will be detrimental to national food security. In the absence of climate adaptation measures, the effects of climate change can be expected to exacerbate the impact of risk events on farm productivity and food security.

References

Akinseye, F. M., Adam, M., Agele, S. O., Hoffmann, M. P., & Traore, P. C. S. (2017). Assessing crop model improvements through comparison of sorghum (sorghum bicolor L. moench) simulation models: A case study of West African varieties. *Field Crops Research, 201*, 19–31. [88]

Akpodiete, N., & Lale, N. (2015). Role of physical characteristics of the seed on the stability of resistance of maize varieties to maize weevil (Sitophilus Zeamais Motschulsky). *IOSR Journal of Environmental Science, Toxicology and Food Technology (IOSR-JESTFT), 9*, 60–66. https://doi.org/10.9790/2402-09226066. [85]

Akponikpè, P. B. I., Gérard, B., Michels, K., & Bielders, C. (2010). Use of the APSIM model in long term simulation to support decision making regarding nitrogen management for pearl millet in the Sahel. *European Journal of Agronomy,32*(2), 144–154. https://doi.org/10.1016/j.eja.2009.09.005 [92]

ANAT. (2016). Environnement et Gestion Des Ressources Naturelles. Sénégal. [105]

Araya, A., Prasad, P. V. V., Gowda, P. H., Zambreski, Z., & Ciampitti, I. A. (2021). Management options for midcentury maize (Zea mays L.) in Ethiopia. *Science of the Total Environment, 758*, 143635. https://doi.org/10.1016/j.scitotenv.2020.143635.[96]

Batini, N. (Ed.). (2021). *The economics of sustainable food: Smart policies for health and the planet*. Island Press.

Bationo, A., Mokwunye, U., Vlek, P. L. G., Koala, S., & Shapiro, B. I. (2003). Soil fertility management for sustainable land use in the West African Sudano-Sahelian zone. In M. P. Gichuru et al. (Eds.), *Soil fertility management in Africa: A regional perspective*. Academy of Science Publishers (ASP) and Tropical Soil Biology and Fertility of CIAT. [93]

Cairns, J. E., Sonder, K., Zaidi, P. H., Verhulst, N., Mahuku, G., Babu, R., Nair, S. K., Das, B., Govaerts, B., Vinayan, M. T., Rashid, Z., Noor, J. J., Devi, P., San Vicente, F., & Prasanna, B. M. (2012). Maize production in a changing climate: impacts, adaptation, and mitigation strategies. In D. Sparks (Ed.), *Advances in agronomy* (Vol. 114, pp. 1–58). Academic Press. [83]

CILSS. (2016). Les Paysages de l'Afrique de l'Ouest: Une Fenêtre sur un Monde en Pleine Évolution. U.S. Geological Survey EROS, 47914 252nd St, Garretson, SD 57030, United States. [101]

Diedhiou, A., Bichet, A., Wartenburger, R., Seneviratne, S. I., Rowell, D. P., Sylla, M. B., Diallo, I., Todzo, S., Touré, N. E., Camara, M., Ngatchah, B. N., Kane, N. A., Tall, L., & Affholder, F. (2018). Changes in climate extremes over West and Central Africa at 1.5 and 2 °C global warming. *Environmental Research Letters, 13*, 065020. https://doi.org/10.1088/1748-9326/aac3e5 [73]

FAO. (1993). *Soil tillage in Africa: Needs and challenges*. FAO soils bulletin 69. FAO. [95]

FAO (Food and Agriculture Organization of the United Nations). (2023). FAOSTAT. Retrieved December 15, 2023, from http://faostat.fao.org [18]

Gano, B., Dembele, J. S. B., Tovignan, T. K., Sine, B., Vadez, V., Diouf D., & Audebert, A. (2021). Adaptation responses to early drought stress of West Africa sorghum varieties. *Agronomy, 11*(443). [87]

Gérardeaux, E., Falconnier, G., Gozé, E., et al. (2021). Adapting rainfed rice to climate change: A case study in Senegal. *Agronomy for Sustainable Development,41*, 57. https://doi.org/10.1007/s13593-021-00710-2 [78]

Glatzel, K. (2018). What can we learn from rising rice production in Senegal? IFPRI, Dakar, Senegal (pp. Issue Post). Retrieved June 29, 2021, from https://www.ifpri.org/blog/what-can-we-learn-rising-rice-production-senegal [76]

Hachigonta, S., Nelson, G. C., Thomas, T. S., & Sibanda, L. M. (Eds.). (2013). *Southern African Agriculture and climate change: A comprehensive analysis*. International Food Policy Research Institute. [75]

Hasegawa, T., Li, T., Yin, X., Zhu, Y., Boote, K., Baker, J., Bregaglio, S., Buis, S., Confalonieri, R., Fugice, J., Fumoto, T., Gaydon, D., Kumar, S. N., Lafarge, T., Marcaida Iii, M., Masutomi, Y., Nakagawa, H., Oriol, P., Ruget, F., Singh, U., et al. (2017). Causes of variation among rice models in yield response to CO_2 examined with free-air CO_2 enrichment and growth chamber experiments. *Scientific Reports,7*, 14858. https://doi.org/10.1038/s41598-017-13582-y [79]

Haussman, B. I. G., Rattunde, H. F., Weltzien-Rattunde, E., Traoré, P. S. C., vom Brocke, K., & Parzies, H. K. (2012). Breeding strategies for adaptation of pearl millet and sorghum to climate variability and change in West Africa. *Journal of Agronomy and Crop Science, 198*, 327–339. [89]

IED. (2015). Agridape: Revue Sur l'agriculture Durable À Faibles Apports Externes. Information centre for Low External Input Agriculture. *AGRIDAP, 36*. [102]

IRD. (2016). Etude Prospective En Soutien à La Programmation Européenne Conjointe: Rapport Final [Etat Des Connaissances Scientifiques Sur Les Grands Enjeux Du Développement, Scénarios d' évolution Au Sénégal]. Senegal. Retrieved November 18, 2020, from https://horizon.documentation.ird.fr/exldoc/pleins_textes/divers16-06/010067188.pdf [103]

Jayawardena, D. M., Heckathorn, S. A., Bista, D. R., Mishra, S., Boldt, J. K., & Krause, C. R. (2017). Elevated CO_2 plus chronic warming reduces nitrogen uptake and levels or activities of nitrogen-uptake and -assimilatory proteins in tomato roots. *Physiology Plant, 159* (354–365). https://doi.org/10.1111/ppl.12532. [98]

Jibrin, J. M. (2012). Simulating planting date and cultivar effects on dryland maize production using CERES-maize model. *African Journal of Agricultural Research,7*, 5530–5536. https://doi.org/10.5897/AJAR12.1303. [84]

Jones, A., Breuning-Madsen, H., Brossard, M., Dampha, A., Deckers, J., Dewitte, O., Gallali, T., Hallett, S., Jones, R., Kilasara, M., Le Roux, P., Micheli, E., Montanarella, L., Spaargaren, O., Thiombiano, L., Van Ranst, E., Yemefack, M., & Zougmore R., (Eds.). (2013). Soil Atlas of Africa. European Commission, Publications Office of the European Union (p. 176). [94]

Koudahe, K., Djaman, K., Bodian, A., Irmak, S., Sall, M., Diop, L., Balde, A. B., & Rudnick, D. R. (2017). Trend analysis in rainfall, reference evapotranspiration and aridity index in southern Senegal: Adaptation to the vulnerability of rainfed rice cultivation to climate change. *Atmospheric Climate Science,07*(04), 20–495. https://doi.org/10.4236/acs.2017.74035 [77]

Lobell, D. B., & Burke, M. B. (2010). On the use of statistical models to predict crop yield responses to climate change. *Agricultural and Forest Meteorology, 150*, 1443–1452. [81]

Ministry of Environment and Durable Development (MEDD). (2018a). National report on land degradation neutrality (LDN) under the United Nations Convention to Combat Desertification. Senegal. Retrieved from https://knowledge.unccd.int/sites/default/files/inline-files/senegal-ldn-country-report.pdf

Melillo, J. M., Butler, S., Johnson, J., Mohan, J., Steudler, P., Lux, H., Burrows, E., Bowles, F., Smith, R., Scott, L., Vario, C., Hill, T., Burton, A., Zhou, Y.-M., & Tang, J. (2011). Soil warming, carbon–nitrogen interactions, and forest carbon budgets. *Proceedings of the National Academy of Sciences of the United States of America, 108* (23), 9508–9512. [99]

Mora, O., & De Lattre, M. (2018). In C. Le Mouël, M. De Lattre-Gasquet, O. Mora (Eds.), *Land use and food security in 2050: a narrow road. Versailles: Ed. Quae* (pp. 20–27). [106]

Parent, B., & Tardieu, F. (2014). Can current crop models be used in the phenotyping era for predicting the genetic variability of yield of plants subjected to drought or high temperature? *Journal of Experimental Botany, 65*(21), 6179–6189. [91]

Robinson, S., et al. (2015). Climate change adaptation in agriculture: Ex ante analysis of promising and alternative crop technologies using DSSAT and IMPACT. IFPRI Discussion Paper 1469. International Food Policy Research Institute (IFPRI). http://ebrary.ifpri.org/cdm/ref/collection/p15738coll2/id/129694 [80]

Rosegrant, M. W., Cline, S. A., Li, W., Sulser, T. B., & Valmonte-Santos, R. A. (2005). Looking ahead: Long-term prospects for Africa's Agricultural Development and Food Security. 2020 Discussion Paper 41. International Food Policy Research Institute. [74]

Rosegrant, M. W., Koo, J., Cenacchi, N., Ringler, C., Robertson, R., Fisher, M., Cox, C., Garrett, K., Perez, N. D., & Sabbagh, P. (2014). Food security in a World of natural resource scarcity—The role of agricultural technologies. International Food Policy Research Institute, Washington, DC

Stone, P. (2001). The effects of heat stress on cereal yield and quality. In A. S. Basara (Ed.), *Crop responses and adaptations to temperature stress* (pp. 243–291). Food Products Press. [82]

Sultan, B., Guan, K., Kouressy, M., Biasutti, M., Piani, C., Hammer, G. L., McLean, G., & Lobell, D. B. (2013). Robust features of future climate change impacts on sorghum yields in West Africa. *Environmental Research Letters, 9*, 104006. [90]

Tappan, G. G., Cushing, W. M., Cotillon, S. E., Mathis, M. L., Hutchinson, J. A., & Dalsted, K. J. (2016). West Africa land use and land cover time series. USGS 4. Retrieved January 16, 2020, from https://pubs.er.usgs.gov/publication/fs20173004 [100]

Taylor, J. R. N. (2015). Overview: Importance of Sorghum in Africa. Afripro: Workshop on the proteins of sorghum and millets: Enhancing nutritional and functional properties for Africa. *Pretoria, 2*(4). [118]; Tine, R. C., Faye, B., Ndour, C. T., Sylla, K., Sow, D., Ndiaye, M., Ndiaye, J. L., Magnussen, P., Alifrangis, M., Bygbjerg, I. C., et al. (2013). Parasitic infections among children under five years in Senegal: Prevalence and effect on Anaemia and nutritional status. *ISRN Parasitology, 2013*, 272701. http://libaccess.mcmaster.ca/login?, http://ovidsp.ovid.com/ovidweb.cgi?T=JS&CSC=Y&NEWS=N&PAGE=fulltext&D=prem&AN=27335851 [86]

Walker, T. S., Alwang, J., Alene, A., Ndjuenga, J., Labarta, R., Yigezu, Y., Diagne, A., Andrade, R., Andriatsitohaina, R. M., De Groote, H., & Mausch, K. (2015). Chapter 19 varietal adoption, outcomes, and impact. In T. S. Walker & J. Alwang (Eds.), *crop improvement, adoption, and impact of improved varieties in food crops in Sub-Saharan Africa* (p. 388). [68]

World Bank. (2008). Senegal—Country environmental analysis. Report No. 48804-SN. Sustainable Development Department Africa Region. United States. Retrieved November 18, 2020, from https://openknowledge.worldbank.org/handle/10986/7773 [104]

Chapter 4
Resulting Impact of Climate Change on Child Malnutrition

4.1 Introduction

According to the Food and Agriculture Organization (FAO) in 2022, around 10% of humanity was suffering from hunger and 35 million more people were pushed into hunger between 2020 and 2022. Add climate change to these statistics, and the situation becomes even more urgent. What is more, half of humanity is chronically malnourished, and most of the rest is either overweight or obese, because the food we produce is less and less nutritious and the way we distribute it globally is grossly insufficient and vulnerable to disruptions (Batini, 2021). A recent report by the International Food Policy Research Institute (IFPRI) and our simulated results in the previous chapters warn that unchecked climate change will have major negative effects on agricultural productivity, with yield declines for the most important crops and additional price increases for the world's staples—rice, wheat, maize, and soybeans (Nelson et al., 2009). It also suggests that there will be 20% more malnourished children in 2050 due to climate change. Moreover, the International Assessment of Agricultural Knowledge, Science and Technology for Development (IAASTD, 2009) argues that climate change, coincident with increasing demand for food, feed, fiber, and fuel, has the potential to irreversibly damage the natural resource base on which agriculture depends, with significant consequences for food insecurity. Climate change could also significantly constrain economic development in developing countries that rely largely on agriculture (Rosegrant et al., 2008). Agricultural production—including access to food—in many African countries and regions will be severely affected. The area suitable for agriculture, the length of growing seasons and yield potential, particularly along the margins of semi-arid and arid areas, are expected to decrease. This would further adversely affect food security and exacerbate malnutrition. As a consequence of these impacts, climate change is likely to further entrench food insecurity in sub-Saharan Africa. By 2080, about 75% of all people at risk of hunger are estimated to live in this region (Nelson et al., 2009).

M. Toure, *Agricultural Adaptation to Climate Change in Sub-Saharan Africa*, Advances in African Economic, Social and Political Development,
https://doi.org/10.1007/978-3-031-80716-9_4

Despite some development progress in recent years, Senegal is a highly food insecure country: estimates suggest that over 15% of rural households, and over 8% of urban households are food insecure (WFP, 2011, 2012). High rainfall variability at inter-annual and inter-decadal scales, particularly in key production areas (Fall et al., 2006; Mcsweeney et al., 2010), affects agricultural production. Geographically, food insecurity is more prevalent in the southern and central parts of the country, especially in the regions of Ziguinchor, Kolda and Kédougou. The analysis presented by WFP (2011), Fall et al. (2006), and McSweeney et al. (2010) suggest a role for climate in the food insecurity in the center, which is the main region of agricultural production, not in the south. This is because the southern regions are climatologically wetter and less variable. Therefore, the food insecurity that characterizes the south may be the consequence of the long-term conflict that has plagued Casamance (Ziguinchor and Kolda regions), and more generally of its 'landlockedness': the difficult transportation and communication conditions linking the south and east with the rest of the country, especially with the port of Dakar. For these same general reasons, lack of access to markets may also be the major contributor to food insecurity in the eastern parts of the country (Kédougou and Matam). For the central part of the country, often referred to as the 'groundnut basin', because this is where cash-crop exploitation took roots during French colonization, the greater dependence of livelihoods on agriculture, and exposure to rainfall variability justifies a more in-depth exploration of the role of climate in shaping food insecurity.

Looking toward the next 25 years, by how much is Senegal's prevalence of child malnutrition likely to decline by 2050? How fast is the decline likely to take place? Given projected population growth, are the number of malnourished children likely to increase or to decrease? The main contribution of this chapter is to answer these questions based on the projected assessments of future developments in the world food situation. While Sect. 4.2 will explore the climate-nutrition literature in sub-Saharan Africa, Sect. 4.3 will estimate child malnutrition in 2050 in Senegal. For that, we will first discuss food availability, historical trend and projected change in cereal consumption and trade, then estimate the projected number of malnourished children based on three scenarios, which rely on the evolution of the underlying determinants of child malnutrition from 1960 to 2022.

4.2 Stylized Facts

To the question: does the effects of climate change influence malnutrition status of children in sub-Saharan Africa? This section is relevant to highlight and inform the urgent need for action to mitigate the effects of climate change on malnutrition of children in sub-Saharan Africa with evidence from literature. Indeed, the influence of climate change on child malnutrition is evident and documented in scientific research. As shown earlier, the effects of these changes may influence land degradation and agricultural productivity. These consequences of climate change severely affect low-income countries in sub-Saharan Africa whose economy depends mainly on agriculture. This reduces food availability and accessibility. These pathways indirectly affect child nutrition and predispose children to malnutrition, most notably stunting, wasting, and undernutrition. Extreme weather events such as prolonged droughts and flooding reduce the availability of food in affected geographical areas thereby exposing millions of children to malnutrition in Sub-Saharan Africa (Rossati, 2017). Furthermore in 2020, 149.2 million children suffered from stunting globally, out of these two out of five live in Africa (41%) (Baker & Anttila-Hughes, 2020).

A study of climatic variability and child weight in 18 African countries[1] indicated that both increased temperatures and low precipitations are associated with reduced weight and height of children (Smith & Haddad, 2020). In this study, the authors estimated a series of multivariate regression models to address two overarching objectives. First, they estimated the association between recent exposure to climatic variability and children's weight, measured both continuously and as a binary indicator of low weight (i.e., wasting). Second, they tested for systematic variation in these effects across socioeconomic groups that they expect to be differentially vulnerable to changing environmental conditions. They focused on this second set of analyses on wasting since the health and social costs of such acute malnutrition are high. Their analyses revealed that exposure to high temperatures was associated with reductions in WHZ (weight-for-length/height z-score) and corresponding increases in the prevalence of wasting. Furthermore, the (IPCC, 2022) report in Climate Change Impacts, Adaptation, and Vulnerability mentioned that climate change and global warming at 2 °C in the midterm undermined food security and nutritional diet in sub-Saharan Africa. This will exacerbate the pressure on food security risk leading to malnutrition and micronutrient deficiencies in sub-Saharan Africa.

Another study, that characterized the contribution of high temperatures to child undernourishment in sub-Saharan Africa, estimated that by the year 2100, child wasting will increase by 37% in Africa due to high temperatures which are harmful to child weight (Baker & Anttila-Hughes, 2020). The authors first identified a strong negative relationship between child weight and average temperature across regions. They then exploited variation in weather conditions to statistically identify the effects of increased temperatures over multiple time scales on child nutrition. Increased temperatures in the month of survey, year leading up to survey and child lifetime led to meaningful declines in acute measures of child nutrition. They found that the lifetime-scale effects explain most of the region-level negative relationship between weight and temperature, indicating that high temperatures may be a constraint on child nutrition. Furthermore, the undernourished population in sub-Saharan Africa is expected to increase due to climate change compared to a future without climate change by 2050. Severe stunting is estimated to increase between 30 and 50% (Lloyd et al., 2011). Therefore, Africa is not on course to reach the UN SDG 2 Zero hunger by 2030 from a predictive model. Child malnutrition will increase by 20% due to climate change and other determinants by 2050 (Atukunda et al., 2021). Food

[1] Benin, Burkina Faso, Cameroon, Côte d'Ivoire, Ethiopia, Ghana, Guinea, Kenya, Madagascar, Malawi, Mali, Niger, Nigeria, Rwanda, Tanzania, Uganda, Zambia, Zimbabwe.

production shortfalls caused by climate variability may increase food prices severely affecting poor households in sub-Saharan Africa. This will have an effect on food affordability and food intake causing malnutrition and micronutrients deficiency in children (Tirado et al., 2015).

Though children are the least responsible for climate change, they will bear the greatest burden of its impact. From droughts to floods to wildfires to deforestation, climate change and environmental degradation is a direct threat to a child's ability to survive, grow, learn, and thrive. The world is suffering a triple burden of child malnutrition, combining undernutrition, micronutrient deficiencies, and overweight and obesity. 144 million children under 5 are stunted, 47 million children under 5 are wasted, 340 million children under 5 are micronutrient deficient, and 38 million children under 5 are overweight (UNICEF, 2019). These challenges are only exacerbated by the compounding effects of climate change and environmental degradation.

4.3 Child Malnutrition in Senegal: A Cross-Country Regression Analysis

4.3.1 Food Availability

The ability of the agricultural sector and food security systems to respond to climatic shocks also depends on trade in cereal products. Empirical evidence has shown that the negative impacts of climate change are dampened when trade patterns adjust (Costinot et al., 2016). Intra-African and international cereals trade is important, because trade has the potential to help reduce the effect of climate change by delivering food from surplus areas to countries experiencing food productivity declines (Duchin & Julia, 2007; Govereh, 2007). To achieve this, there is a need to promote and facilitate freer movement of cereals. Free trade allows nations to fully exploit their comparative advantage and the existence of barriers to trade exacerbate climate change effects by reducing the responsiveness of consumers and producers to incentives. This points to the importance of trade policies in SSA if they are to respond effectively to changes in climate, among other adaptation strategies that include irrigation and the adoption of improved varieties. Empirical evidence has shown that trade costs are high in SSA and have a negative effect on food prices, agricultural income, and welfare (Porteous, 2016). To restrict the number of agricultural products in the analysis, the study focuses on the trade of the following cereals: rice, maize, and sorghum in Senegal, as it forms the major share of food consumption goods.

4.3.1.1 Cereal Historical Trade

Between 2000 and 2016, the share of cereal exports in total trade was low for the majority of countries in sub-Saharan Africa (SSA). Yet, Senegal had the highest share of cereal exports (Porteous, 2016). The share of cereal exports had been increasing in the country. The share of exports increased from 0.01% in 2000 to 6% in 2005, and dropped to 4% in 2016. Agricultural trade is of critical importance for Senegal as well as for many SSA countries. Improvements in agricultural trade in the region are likely to be associated with economic growth and poverty reduction, as most poor people in the region are dependent on agricultural production for income generation and food security. SSA countries are known for being chronically food insecure and highly dependent on food aid. Agricultural trade is important for linking food surplus zones with food deficit regions, thereby increasing food security in deficit regions through food availability and accessibility (Taylor, 2015). This will be particularly important in the face of extreme events such as frequent occurrences of droughts and flooding as a result of climate change. In addition to increasing the resilience of food security systems to cater for changes in climate through delivering food to deficit regions, agricultural trade will also reduce price volatility, improve producer incentives, and increase agricultural growth. As shown in Fig. 4.1, Senegal has increasingly become a net importer of food and agricultural products in the past two decades. These constituted mainly imports of cereals and livestock products, which are important for improving food security and the nutritional status of individuals. Climate shocks may affect the importation of these products, as they are likely to be associated with the potential of countries to increase their import of food through changes in comparative advantage and income. About 30% of agricultural imports into SSA come from other developing regions such as Asia, with Thailand, Malaysia, India, and Indonesia being the top countries (Duchin & Julia, 2007). The poor performance of intra-regional trade can be attributed to several factors, most importantly trade policies, complex transit procedures, poor institutional environments and high transport costs. Josling (2008) and Porteous (2016) note that SSA countries face much higher trade-related costs than other countries in getting their products to international markets. This is likely to exacerbate the impact of climate shocks. For example, high trade costs have been shown to reduce the adoption of agricultural technologies that may be important for adapting to climate change shocks (Porteous, 2016).

4.3.1.2 Predicted Change in Cereal Trade

In a recent study called 'Climate Change and Agricultural Trade Sub-Saharan Africa', the average cereals productivity in the whole SSA is expected to decrease by 2.32%, with wheat, maize, and rice productivity decreasing by 22.7%, 11.7%, and 3.9%, respectively, whereas productivity of crops such as millet and sorghum is likely to increase (Mahofa, 2022). This is because millet and sorghum are drought-resistant crops and are likely to grow well in dry and high temperature conditions that are a result of climate change. The results show that the potential of most SSA countries to export maize will decline. In West Africa, the potential export of maize will decline by 6% in Senegal. Furthermore, Senegal is expected to experience relatively large

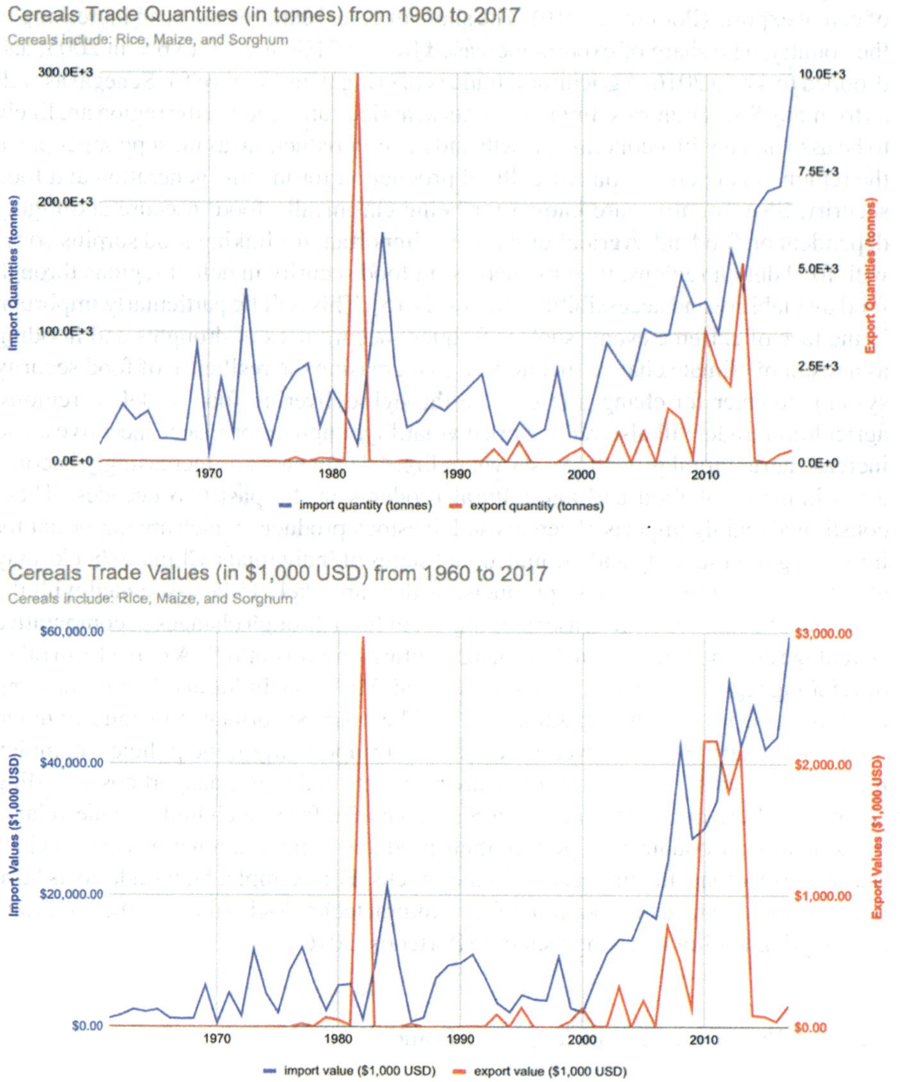

Fig. 4.1 Historical Cereal Trade quantities and values from 1960 to 2020. *Source* Author's computations based on FAOSTAT database

Table 4.1 Percent change in Senegal Trade in 2050s

Crop	Maize (%)		Rice (%)		Sorghum (%)	
	X	M	X	M	X	M
	–6.23	5.25	–8.85	2.3	–0.82	1.12

Notes X represents exports and M represents import
Source "Climate Change and Agricultural Trade in Sub-Saharan Africa" by Mahofa (2022)

declines in rice exports of 8.9% in the 2050s, as shown in Table 4.1. Mali may import US$2 million worth of rice from Senegal while Senegal's export supply is likely to be US$19.5 million.

4.3.2 Projection of Child Malnutrition in Cereal Trade

4.3.2.1 Historical Trend

The causes of child malnutrition are complex, multidimensional, and interrelated. They range from factors as broad in their impact as political instability and slow economic growth to those as specific in their manifestation as respiratory infection and diarrheal disease. In turn, the implied solutions vary from widespread measures to improve countries' stability and economic performance to efforts to enhance access to sanitation and health services in individual communities. Debates continue to flourish over what the most important causes of malnutrition are and what types of interventions will be most successful in reducing it. An understanding of the most important causes of malnutrition is imperative if the current unacceptably high numbers of malnourished children are to be reduced. Even though malnutrition is associated with both undernutrition and overnutrition. In our study, we use the term to refer to cases of undernutrition as measured by underweight rates. A child is considered underweight if the child falls below an anthropometric cutoff of –2 standard deviations below the median value of weight-forage using National Center for Health Statistics/World Health Organization standards.

According to a recent USAID report in 2020 on child malnutrition in Senegal, the prevalence of chronic malnutrition (stunting or low height-for-age) is lower in Senegal than in many other West African countries, but still affects 17% of children under 5 years but reaching its peak of 23% among children 24–35 months (USAID, 2020). Additional studies found that the basic causes of child malnutrition (or stunting) in Senegal included climate shocks including rain, droughts and extreme cold, economic shocks including periods of increasing food prices, rural residence, household size (Diouf et al., 2000), household wealth, and parental education (UNICEF, 2019). These studies were in agreement that children residing in rural areas were at a greater risk of stunting as compared to their urban counterparts. Girls were found to have some gains with catch-up growth when they migrated for a short period of time to an urban area (Dakar) from a rural area (Niakhar). A study examining migration patterns found that during periods of heat waves and droughts, individuals from areas where there is acute malnutrition are more likely to migrate compared to those from areas without chronic malnutrition (UNICEF, 2019).

4.3.2.2 Methodology

The precise relationship that we used to project the percentage of malnourished children is based on a cross-country regression relationship of Smith & Haddad (2000). The conceptual framework underlying their study is adapted from the United Nations Children's Funds framework for the Causes of Child Malnutrition (UNICEF, 1990, 1994; Walker et al., 2015; Willmott, 1982). The framework recognizes three levels of causality corresponding to immediate, underlying, and basic determinants of child nutritional status.

The immediate determinants of child nutritional status manifest themselves at the level of the individual human being. They are dietary intake (energy, protein, fat, and micronutrients) and health status. These factors themselves are interdependent. A child with inadequate dietary intake is more susceptible to disease. In turn, disease suppresses appetite, inhibits the absorption of nutrients in food, and competes for a child's energy. Dietary intake must be adequate in quantity and in quality, and nutrients must be consumed in appropriate combinations for the human body to be able to absorb them.

The immediate determinants of child nutritional status are, in turn, influenced by three underlying determinants manifesting themselves at the household level. These are food security, adequate care for mothers and children, and a proper health environment, including access to health services. Associated with each is a set of resources necessary for their achievement. Food security is achieved when a person has access to enough food for an active and healthy life (World Bank, 1986). Care, the second underlying determinant, is the provision in households and communities of 'time, attention, and support to meet the physical, mental, and social needs of the growing child and other household members' (ICN, 1992).

Finally, the underlying determinants of child nutrition (and poverty) are, in turn, influenced by basic determinants. The basic determinants include the potential resources available to a country or community, which are limited by the natural environment, access to technology, and the quality of human resources. Political, economic, cultural, and social factors affect the utilization of these potential resources and how they are translated into resources for food security, care and health environments and services.

With a particular focus on the underlying determinants of child malnutrition, the following variables are considered as shown in Eqs. (C.10) and (C.11): (i) national food availability, (ii) women's education, and (iii) women's status relative to men, indeed, the education level of women—the main caretakers of children—may have several potential positive effects on the quality of care, and (iv) access to safe water as a proxy for health environment and services.

$$\begin{aligned}\Delta_{t,2020}\text{MAL} = & -25.24 \times ln\left(\frac{\text{KCAL}_t}{\text{KCAL}_{2020}}\right)\\ & - 71.76 \times \Delta_{t,2020}\text{LFEXPRAT}\\ & - 0.22 \times \Delta_{t,2020}\text{SCH}\\ & - 0.08 \times \Delta_{t,2020}\text{WATER}\end{aligned} \tag{4.1}$$

where

MAL	is the percentage of malnourished children
KCAL	is the per capita kilocalorie availability
LFEXPRAT	is the ration of female to male life expectancy at birth
SCH	is the total female enrollment in secondary education (any age group) as percentage of the female age-group corresponding to national regulation for secondary education
WATER	is the percentage of population with access to safe water
$\Delta_{t,2020}$	is the difference between the variable values at time t and the base year t_{2020}.

$$\text{NMAL}_t = \text{MAL}_t \times \text{POP}_t \tag{4.2}$$

where

NMAL	is the number of malnourished children, and
POP	is the number of children under five years old in the population.

Data on the percentage of malnourished children (MAL are taken from the World Development Indicators). Other data sources include the FAOSTAT database and the UNESCOSTAT database. Appendix E shows the historical trend and projected estimations of the variables used in Eqs. (C.10) and (C.11) and a detailed description of each variable source and relevance to this study.

4.3.2.3 Simulated Results

Simulated results in the cereal availability and consumption are tabulated in Appendix E. In order to compute the projected number of malnourished children in Senegal in 2050, we applied Eqs. (C.10) and (C.11) in three scenarios based on the projected evolution of the underlying determinants of child malnutrition over the period 2020–2050. These are (i) a status quo, or 'do nothing different' scenario, (ii) a pessimistic scenario, and (iii) an optimistic scenario, which are the same scenarios defined by Smith and Haddad (2000). The assumptions underlying each scenario are as follows.

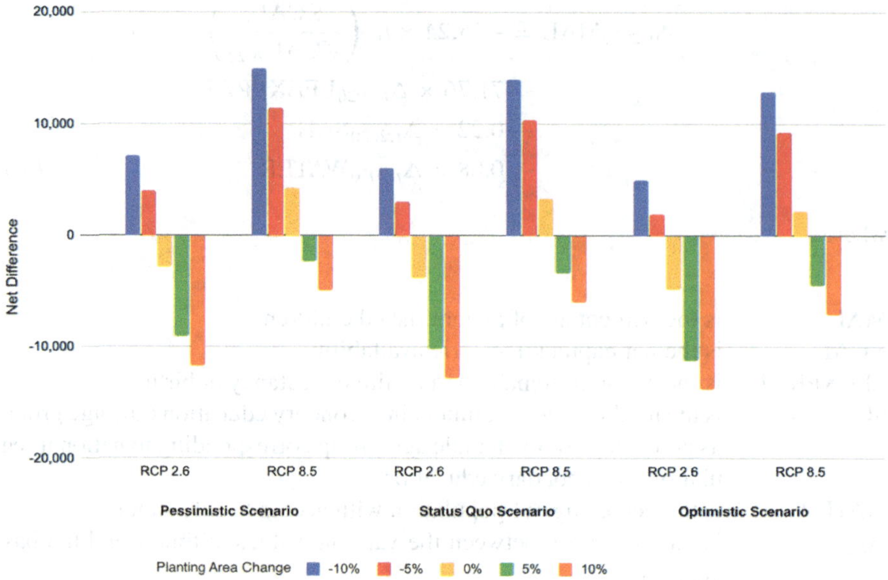

Fig. 4.2 Net difference estimated number of Malnourished Children in 2050. *Source* Author's computations

In the status quo scenario, safe water access, female secondary enrollments, and the female-to-male life expectancy ratio improve at the same rates they improved over 2010 and 2020. At these rates, as shown in Appendix E, an increase of 17% of the Senegalese population would have access to safe water by the year 2050. Female secondary enrollment would increase from 34 to 51% in 2050. The female-to-male life expectancy ratio would rise to 1.07. Given that the number of malnourished children is 400,000 in 2020 (FAOSTAT database), our simulated results, as shown in Fig. 4.2 and Table 4.2, the expected number of malnourished children varies from 387,278 to 413,933. The minimum value, which represents a 3% decrease in the number of malnourished children, is observed at a 'low emission' climate change scenario with a 10% increase in planting area change. While the highest number of malnourished children is synonymous with a 3.5% increase in the number of malnourished children, is observed at a 'high emission' climate change scenario with a 10% decrease in planting area change.

Under the pessimistic scenario, the rate of improvement in the nonfood underlying determinants is assumed to decline by 25%. Under this circumstance, access to safe water rises to only 30% (compared to 26% in 2020 as shown in Appendix E), female secondary enrollment rises to 64%, and the female-to-male life expectancy ratio to 1.08. This projection is based on a 'low-investment/slow growth' scenario in which international donors eliminate public investment in national agricultural research systems and extension services in developing countries and phase out direct core funding of international agricultural research centers. With 400,000 malnour-

Table 4.2 Estimated number of Malnourished children in Senegal in 2050

	Pessimistic scenario		Status quo scenario		Optimistic scenario	
Percentage change in planting area (%)	RCP 2.6	RCP 8.5	RCP 2.6	RCP 8.5	RCP 2.6	RCP 8.5
–10	407,084	414,980	406,038	413,933	404,992	412,887
–5	404,022	411,393	402,976	410,347	401,930	409,300
0	397,265	404,310	396,219	403,264	395,173	402,218
5	390,935	397,692	389,889	396,646	388,843	395,600
10	388,324	395,049	387,278	394,003	386,231	392,957

Source Author's computations

ished children in 2020 (based on the FAOSTAT database), the expected number of malnourished children, under this pessimistic scenario, is expected to vary between 388,324 and 414,980. The minimum value, which represents 11,676 more malnourished children compared to 2020 and 1,046 more malnourished children compared to the 'status quo' scenario, is observed at a 'low emission' climate change scenario with a 10% increase in planting area change. While the highest number of malnourished children is synonymous with an increase in the number of malnourished children of 14,980 compared to 2020 and 1,047 more malnourished children compared to the 'status quo' scenario, is observed at a 'high emission' climate change scenario with a 10% decrease in planting area change.

Under the optimistic scenario, the rate of improvement in the nonfood underlying determinants is assumed to increase by 25%. Under this circumstance, access to safe water rises to only 32% (compared to 26% in 2020 as shown in Appendix E), female secondary rises to 73%, and the female-to-male life expectancy ratio to 1.09. Since the FAOSTAT database assumes that there are 400,000 malnourished children in 2020, the expected number of malnourished children, under this optimistic scenario, is expected to vary between 386,231 and 412,887. The minimum value represents 13,769 less malnourished children compared to 2020 and 1.047 less malnourished children compared to the 'status quo' scenario and is observed at a 'low emission' climate change scenario with a 10% increase in planting area change. While the highest number of malnourished children is synonymous with an increase in the number of malnourished children of 12,887 compared to 2020 and 1,047 less malnourished children compared to the 'status quo' scenario, is observed at a 'high emission' climate change scenario with a 10% decrease in planting area change.

In line with our findings, the Intergovernmental Panel on Climate Change (IPCC) report published in 2022 argues that climate change is projected to increase malnutrition through reduced nutritional quality, access to balanced food and inequality. Increased CO_2 concentrations (highlighted in our analysis with 'low emission' scenario or RCP 2.6 and 'high emission scenario' or RCP 8.5) promote crop growth and yield but reduce the density of important nutrients in some crops with projected increases in undernutrition and micronutrient deficiency, particularly in countries that currently have high levels of nutrient deficiency and regions with low access to diverse foods. Moreover, climate change will increasingly add pressure on food production systems, undermining food security. With every increment of warming, exposure to climate hazards will grow substantially, and adverse impacts on all food sectors will become prevalent, further stressing food security. Regional disparity in risks to food security will grow with warming levels, increasing poverty traps, particularly in regions characterized by a high level of human vulnerability.

Our findings underline the need for continued attention to climate impacts on agriculture, livelihoods, and health, which together determine children's nutritional status. Rather than assume that climate affects nutrition solely through changes in agricultural production and food availability, research must account for the multiple pathways through which this relationship operates. Future work should draw on the rich, longitudinal datasets that are available in many countries within the region to assess the mechanisms that have been hypothesized but largely untested. Also, more evidence is needed to design and evaluate interventions that can enhance resilience to climatic variability as it relates to children's health and nutrition. There has been some attention to the ways in which development interventions (e.g., cash transfers) may moderate the effects of climatic variability on population processes in Africa (Mueller et al., 2020), but overall very little evidence is available to design interventions that mitigate the demographic and health impacts of climatic change. The large number of governmental (e.g., Ethiopia's Productive Safety Net Program) and non-governmental (e.g., GiveDirectly cash transfers) social protection programs being implemented across sub-Saharan Africa provide ample opportunity to conduct such assessments. This line of future research must also give significant attention to the causal pathways through which climate effects operate, since policy interventions must be tailored accordingly. Given both the short- and long-run consequences of child malnutrition, continued efforts are needed to refine our understanding of whether, how, and why such outcomes are influenced by environmental change. Ultimately, research can better inform policies and programs that improve nutritional outcomes in the face of ongoing climatic variability.

References

Atukunda, P., Eide, W. B., Kardel, K. R., Iversen, P. O., & Westerberg, A. C. (2021). Unlocking the potential for achievement of the UN Sustainable Development Goal 2—'Zero Hunger'—In Africa: Targets, strategies, synergies and challenges. *Food & Nutrition Research, 65*. https://doi.org/10.29219/fnr.v65.7686 [115]

Baker, R. E., & Anttila-Hughes, J. (2020). Characterizing the contribution of high temperatures to child undernourishment in Sub-Saharan Africa. *Scientific Reports, 10*(1), 18796. https://doi.org/10.1038/s41598-020-74942-9 [111]

Batini, N. (Ed.). (2021). *The economics of sustainable food: Smart policies for health and the planet*. Island Press.

Costinot, A., Donaldson, D., & Smith, C. (2016). Evolving comparative advantage and the impact of climate change in agricultural markets: Evidence from 1.7 million fields around the world. *Journal of Political Economy, 124*(1), 205–47. [118]

Diouf, S., Diallo, A., Camara, B., Diagne, I., Sy Signate, H., Sarr, M., Fall, M. (2000). Parasitoses Intestinales De L'Enfant en Zone Rurale Sénégalaise (Khombole). *Dakar Medical, 47*(5). Retrieved from http://www.academia.edu/32251225/Parasitoses_Intestinales_De_LEnfant_en_Zone_Rurale_Senegalaise_Khombole [125]

Duchin, F., & Juliá, R. (2007). World trade as the adjustment mechanism of agriculture to climate change. *Climate Change, 82*(3), 393–409. [120]

Fall, S., Semazzi, F. H. M., Niyogi, D. D. S., Anyah, R. O., & Bowden, J. (2006) The spatio-temporal climate variability over Senegal and its relationship to global climate. *International Journal of Climate, 26*, 2057–2076. [24]

Govereh, J. (2007). Regional trade in food staples: Using trade policy to improve farmer incentives and food security. In *Paper presented at the Conference on Strengthening and Widening Markets and Overcoming Supply Side Constraints for African Agriculture*. [119]

IAASTD. (2009). Agriculture at a crossroads. International Assessment of Agricultural Knowledge, Science and Technology for Development. Island Press [108]

ICN (International Conference on Nutrition). (1992). Caring for the socio-economically deprived and nutritionally vulnerable. Major Issues for Nutrition Strategies Theme Paper # 3. ICN/92/INF/7. Food and Agriculture Organization of the United Nations/World Health Organization. [132]

IPCC. (2022). Climate Change 2022: Impacts, Adaptation and Vulnerability. Contribution of Working Group II to the Sixth Assessment Report of the Intergovernmental Panel on Climate Change.; Pörtner, H.-O., Roberts, D.C., Tignor, M., Poloczanska, E. S., Mintenbeck, K., Alegría, A., Craig, M., Langsdorf, S., Löschke, S., Möller, V., Okem, A., & Rama, B. (Eds.). Cambridge University Press (3056 pp.). https://doi.org/10.1017/9781009325844 [134] [113]

Josling, T. (2008). World Bank work programme on regional integration and agricultural trade policy framework document. The World Bank. [122]

Lloyd, S. J., Kovats, R. S., & Chalabi, Z. (2011). Climate change, crop yields, and undernutrition: Development of a model to quantify the impact of climate scenarios on child undernutrition. *Environmental Health Perspectives,119*(12), 1817–1823. https://doi.org/10.1289/ehp.1003311 [114]

Mahofa, G. (2022). Climate Change and Agricultural Trade in Sub-Saharan Africa. Research Paper 499, African Economic Research Consortium. Retrieved May, 12, 2023, from http://publication.aercafricalibrary.org/ [123]

Mcsweeney, C., New, M., Lizcano, G., & Lu, X. (2010). The UNDP Climate Change Country Profiles: Improving the accessibility of observed and projected climate information for studies of climate change in developing countries. *Bulletin of the American Meteorological Society,91*, 157–166. https://doi.org/10.1175/2009BAMS2826.1 [8]

Mueller, V., Sheriff, G., Dou, X., & Gray, C. (2020). Temporary migration and climate variation in eastern Africa. *World Development,126*, 104704. https://doi.org/10.1016/j.worlddev.2019.104704 [133]

Nelson, G. C., Rosegrant, M. W., Koo, J., Robertson, R., Sulser, T., Zhu, T., Ringler, C., Msangi, S., Palazzo, A., Batka, M., Magalhaes, M., Valmonte-Santos, R., Ewing, M., & Lee, D. (2009). *Climate change. Impact on agriculture and costs of adaptation*. IFPRI [107]

Porteous, O. C. (2016). Essays on agricultural trade in sub-Saharan Africa. Unpublished Ph.D. thesis, University of Berkeley, Department of Agricultural and Resource Economics. [121]

Rosegrant, M. W., Ewing, M., Yohe, G., Burton, I., Huq, S., & Valmonte-Santos, R. (2008). Climate change and agriculture: Threats and opportunities. GTZ on behalf of Federal Ministry for Economic Cooperation and Development, Eschborn. [109]

Rossati, A. (2017). Global warming and its health impact. *The International Journal of Occupational and Environmental Medicine, 8*(1), 7–20. https://doi.org/10.15171/ijoem.2017.963 [110]

Smith, L., & Haddad, L. (2000). Explaining Child Malnutrition in Developing Countries: A cross-country analysis. In *IFPRI Research Report*. International Food Policy Research Institute. [128]

Taylor, J. R. N. (2015). Overview: Importance of Sorghum in Africa. Afripro: Workshop on the proteins of sorghum and millets: Enhancing nutritional and functional properties for Africa. *Pretoria,*

2(4). [118]; Tine, R. C., Faye, B., Ndour, C. T., Sylla, K., Sow, D., Ndiaye, M., Ndiaye, J. L., Magnussen, P., Alifrangis, M., Bygbjerg, I. C., et al. (2013). Parasitic infections among children under five years in Senegal: prevalence and effect on Anaemia and nutritional status. *ISRN Parasitology, 2013*, 272701. Retrieved from http://libaccess.mcmaster.ca/login?url. http://ovidsp.ovid.com/ovidweb.cgi?T=JS&CSC=Y&NEWS=N&PAGE=fulltext&D=prem&AN=27335851 [86]

Thiede, B. C., & Strube, J. (2020). Climate variability and child nutrition: Findings from Sub-Saharan Africa. *Global Environmental ChangeÃ¢â,¬Â¨: Human and Policy Dimensions,65*, 102192. https://doi.org/10.1016/j.gloenvcha.2020.102192 [112]

Tirado, M. C., Hunnes, D., Cohen, M. J., & Lartey, A. (2015). Climate Change and nutrition in Africa. *Journal of Hunger & Environmental Nutrition,10*(1), 22–46. https://doi.org/10.1080/19320248.2014.908447 [116]

UNICEF. (2019). The State of the World's Children (SOWC) Report 2019. UNICEF. [117]

UNICEF (United Nations Children's Fund). (1990–1992). The State of the World's Children 1990. [129]

UNICEF (United Nations Children's Fund). (1994–1998). The State of the World's Children 1994. [130]

USAID. (2020). Nutrition profile: Senegal. Retrieved from https://2017-2020.usaid.gov/sites/default/files/documents/1864/Senegal-Nutrition-Profile-Mar2018-508.pdf [124]

Walker, T. S., Alwang, J., Alene, A., Ndjuenga, J., Labarta, R., Yigezu, Y., Diagne, A., Andrade, R., Andriatsitohaina, R. M., De Groote, H., & Mausch, K. (2015). Chapter 19 varietal adoption, outcomes, and impact. In T. S. Walker & J. Alwang (Eds.), *Crop improvement, adoption, and impact of improved varieties in food crops in sub-Saharan Africa* (p. 388). [68]

Willmott, C. J. (1982). Some comments on the evaluation of model performance. *The Bulletin of the American Meteorological Society (BAMS), 63*(11), 1309–1313 [70]

World Food Programme (2012) Food security monitoring system (SSSAS–Système de suivi de la sécurité alimentaire au Sénégal). WFP. [15]

World Bank. (1986). Poverty and Hunger: Issues and Options for Food Security in Developing Countries. [131]

Chapter 5
Policy Issues, Climate Risks, and Financialization of Crops

5.1 Introduction

In order to better assess policy issues and identify those who are being mostly affected by the impact of climate change in cereal production, it is important to discuss the heterogeneity of land ownership. Ninety-five percent of Senegal's farms are smallholdings, with farm size ranging from 1.5 to 5 ha. The remaining 5% of holdings are larger farms. Large national and international agribusinesses have had a presence in Senegal since at least the 1930s. The largest farms have concentrated operations in the Senegal River Valley, the land around Dakar and coastal areas, and in the Casamance, although agribusinesses have also begun to invest in more remote areas, according to the U.S. Agency for International Development.

Since the late 2000s, the land rush on the part of foreign investors has left many Senegalese smallholders by the wayside. In a country where agriculture, the leading employment sector, provides almost 70% of the national labor force with jobs but contributes little to GDP, the State looks kindly on the arrival of private investors such as agro-industry, to foster economic development. CIRAD[1] in partnership with the national land tenure governance observatory in Senegal identified three main categories of land transaction players: (i) Senegalese smallholders, who practice traditional cultivation on inherited land, (ii) local religious, political, or economic elites, who are starting to show an interest in speculative rice or horticultural crop growing, and lastly (iii) overseas investors. According to a CIRAD geographer Jérémy Bourgoin, there are not many (of those investors), but they are the ones with the most impact. That is because they are responsible for converting wooded areas, rangelands, and pastoral zones into agricultural land, with adverse consequences for biodiversity, climate and local communities' livelihoods.

[1] CIRAD is the French agricultural research and cooperation organization working for the sustainable development of tropical and Mediterranean regions. CIRAD and its partners in Senegal have been working for several years to analyze land use dynamics observation structures in Senegal and provide players with tools to boost their impact.

M. Toure, *Agricultural Adaptation to Climate Change in Sub-Saharan Africa*, Advances in African Economic, Social and Political Development,
https://doi.org/10.1007/978-3-031-80716-9_5

Based on this agricultural research organization, 46% of Senegal's total area is arable land, 270 908 ha, which represents 3% of the country's arable land are managed by foreign investors. Also, more than 50% of transactions concern areas of less than 5,000 ha and French and Italian investors account for 55% of transactions. Furthermore, 77% of transactions are concluded to produce agrofuels, food crops concern 28% of projects but 8% of the area under contract, and all the transactions concluded concern land previously used by smallholders. Foreign investments are expected to meet three priority goals set by the current government: developing food self-sufficiency, facilitating food exports and income generation for farmers, and improving basic infrastructure in remote rural and less productive areas.

Land in the national domain cannot be sold, as a result the formal land-sale market in Senegal is limited to privately held freehold land. Freehold land was estimated at 2–3% of total landholdings in 1964. The amount of registered freehold land has increased because some holders of surface rights have converted their rights to freehold rights. However, that number is presumed to be quite small due to the limited number of informal settlements regularized and the high cost of obtaining freehold title. Senegal has an active informal market in land sales and leases. In many rural areas, unexploited rainfed land is available for allocation by rural councils or traditional authorities and thus lacks a market value. However, irrigated land, land identified for development projects, land with access to roads and railways, and land in peri-urban areas have become increasingly valuable and support a vibrant informal land market. Land also has non-economic functions in Senegal, including social functions (e.g., defining membership and status of land users and holders) and religious functions (e.g., mediating between the worlds of the living and the departed), based on the U.S. Agency for International Development.

Originally established in 2009 to address the lack of robust data on Large Scale Land Acquisitions (LSLAs), the Land Matrix Initiative (LMI)[2] is an independent global land monitoring initiative made up of a number of global and regional partners.[3] Its online platform presents detailed large-scale land acquisition (LSLA) transactions in Senegal that have been concluded since the year 2000 and have an intended size of 200 ha or larger. Based on their platform, a total of 270,908 ha are under contract by foreign investors, equaling 3% of the total arable land. Concluded deals are considerably smaller in size than their original intention (obtained 53.8% of the intended size). Of the concluded deals, the large majority are leaseholds (duration of lease between 10 and 99 years). Senegal is only marginally affected by the so-called 'mega deals' because more than 50% of the deals have an area under contract smaller than 5,000 ha. In fact, two-thirds of the concluded deals are operational (start-up phase and in operation). 26% of the deals have been abandoned for various reasons: project no longer viable, bankruptcy of the investor, or community protests. 32% of the area under contract is currently under production (cases with known area under

[2] https://landmatrix.org.

[3] Those partners include CIRAD, the German Institute for Global and Area Studies/Leibniz-Institut für Globale und Regionale Studien (GIGA), the European Commission (EC), and the Swiss Agency for Development and Cooperation (SDC).

production only). What is more, foreign large-scale land acquisitions accelerated since 2004 and peaked in 2010–2011.

After a quiet period (2012–2014), LSLAs have been increasing again recently. An investor from Romania is involved in the largest size under contract. French and Italian investors are engaged in most deals (9 deals out of 19 (55%)). Regional investors also play a role, although to a lesser degree, in land deals in Senegal; there are two deals with involvement by investors from Cameroon and Nigeria. About the dynamics of deals by investor type, private companies are the major type of investors in Senegal, followed by individual entrepreneurs. This being said, the investor type of many investors is not known (no information). Foreign investors do not engage domestic investors; domestic partners hold part ownership in smaller deals only representing 10.5% of the total. Most concluded deals are for production of agrifuel (Jatropha), covering 77% of the area under contract. Food crops concern a large number of projects (28%), however they only account for a small size under contract (8%). Deals for food crops have the highest number of operational cases. Another important point is that all concluded deals, for which former land use is known, have acquired land that was formerly used for smallholder agriculture.

Gender is a critical variable in relation to this important resource in that it helps us to better understand the role of women and men in institutions. Whether these are formal or informal, local or global: besides cultural and social standards, the division of labor is a key factor that determines access and control of land. A survey released in 2010 by the Participatory Geographic Information Systems (P-GIS) for natural resource management and food security in Senegal showed that land ownership in rural households in Senegal is 84.90% nationally. However, the figures vary slightly depending on eco-geographic areas. As expected, the rates of ownership are higher in areas of low population density where land issues seem to be less visible. Thus, eastern Senegal is in the lead with 93.33%, followed closely by Casamance with 92.91% and the forestry/pastoral zone with 84,62%. In contrast, rates are lower in the Valley (80.16%), in the Niayes (80.13%), and in the Groundnut Basin (79.91%), where high densities often involve greater pressure on land. These last two areas are characterized respectively by the increasing implementation of agro-industry and speculation related to the strong demand for land for housing. In the different eco-geographical zones, inheritance, bequests, gifts, and loans emerge as the principal means of access to land. These types of access prior to the law on state land property (assignment) still retain a large social legitimacy. Often, these forms of access (family allowance, legacy) in which men are privileged, serve as reference for local officials who administer the land, because they are well rooted in their socio-mental and socio-cultural universe. Thus, it is largely admitted that access to land is highly dependent on the social status (ethnicity, social class, etc.) and family status (special position that men and women occupy in the household) of individuals. Focusing on the gender of household heads, the results show 94.53% of households with land are headed by a man and 5.47% are headed by women. For households with no land, 87.78% are managed by men, whereas 12.22% are made up of women. There are some cases of individual ownership of land by women. Taking into account these special cases can help better appreciate the role of women within the family and their effective participation in land activities. They are usually widows or divorced women: in any case whose special status turns into de facto heads of household.

Based on this land ownership literature in Senegal, even though agriculture in Senegal is dominated by family-owned farms raising diversified crops and livestock, today, food and agriculture, also known as the agri-food sector, is increasingly being responsible for feeding the planet. Hence, small land owners and women are the most affected by the impact of climate change in cereal production. Section 5.2 will focus on policy issues, recommendations, and their evolution. The role of climate finance to adapt and to mitigate food insecurity will be discussed in Sect. 5.3. This section will wrap up with a discussion of crop financialization in Senegal.

5.2 Policy Issues and Recommendations

5.2.1 Current Hunger Context

Based on the results from previous sections, one can see that climate change impacts are negatively affecting agriculture, the Senegalese food system is failing to address food insecurity and malnutrition in an environmentally sustainable way, and the impacts of climate change on food systems affect everyone, but particularly children. According to the 2017 Brookings Institution report (Ibrahima et al., 2017), Senegalese public authorities have made significant efforts to define strategies to combat food insecurity and malnutrition, as well as provide financing and infrastructure for agricultural policy implementation. Senegal's government agricultural spending is $60.60 per capita, the fourth-highest in sub-Saharan Africa, which has an average agricultural spending of $22.90 per capita. Despite this accomplishment, progress made in achieving Food and Nutrition Security (FNS) is insufficient, as noted above. High expenditures, combined with food insecurity, suggest inefficiencies and weak control of agricultural expenditures. So far, productivity gains fail to match the government's ambitious targets, in part due to institutional inefficiencies embedded in these agricultural support programs. Moreover, they argue that agricultural subsidies do not consistently reach their targeted beneficiaries, and the government may overemphasize spending on input supply services, which have short-term effects on production, but not on long-term growth in the sector. Nevertheless, there have been some noteworthy performances recently in Senegal's FNS programming. Ongoing and recently established initiatives, including interventions related to rice self-sufficiency, child malnutrition (Nutrition Enhancement Program), and conditional cash transfers and emergency contingency planning show promise in addressing the country's priority FNS needs.

Moving forward, how must the government redesign its FNS policies and better target its beneficiaries (i.e., the most vulnerable populations) to increase the efficiency of its operations? What strategies should the government put in place in order to increase agricultural productivity through reallocation of resources toward more targeted investments in infrastructure, research, and human capital for more sustainable gains, and to reduce the high volatility of food production in Senegal and the country's vulnerability to environmental shocks? What is more, the analysis of the state of food and nutrition security in Senegal also shows the need to focus on improving food quality as one of the top priority areas for Senegal. Thus, what policies to adopt to strengthen the quality of food consumption, to promote dietary diversification, and to increase consumption of foods high in calories and protein? Given the nutritional gaps in Senegal relative to other African and developing countries, what efforts should be made to focus on supporting the fight against anemia and reducing the high prevalence of underweight children?

5.2.2 Adaptation Strategies

The 2022 Intergovernmental Panel on Climate Change (IPCC) report highlights a range of options for water- and food-related adaptation in different sociocultural, economic, and geographical contexts, with benefits across several dimensions across regions, including climate risk reduction. Those options include rainwater harvesting, soil moisture conservation, cultivar improvements, community-based adaptation, agricultural diversification, climate services, and adaptive eco-management in fisheries (Ibrahima et al., 2017). Indeed, adaptation strategies to enhance food security vary from farm-level interventions to national policies and international agreements. They cover the following dimensions of food security: availability, access, utilization (food quality and safety), and stability. The report emphasized that adaptation strategies for crop production, as mentioned above, include field and farm-level options such as crop management, livelihood diversification, but also social protection such as crop insurance. The most common field management options are changes in planting schedules, crop varieties, fertilizers, and irrigation. For example, farmers can shift their planting schedules in response to the early or late onset of the rainy season. Moreover, there are new crop insurance schemes that are based on changes in weather patterns.

Additionally, in her book the **Economics of Sustainale Food Smart Policies for Health and Planet**, Dr. Batini highlights that the greening production in less industrialized countries rests on three pillars: healthy diets and nutrition security, minimal environmental impacts, and climate-resilient production systems. About the third pillar, climate resilience, farmers need to adopt climate-resilient strategies. Crop switching, change in planting dates, dissemination of climate information and weather forecasts, and crop insurance are among the approaches that extension activities and policies can include to promote climate resilient agriculture. Indeed, the technological achievements to increase production rank among humanity's greatest achievements. Low food prices and trade supported a dramatic decline in the number of undernourished people, from approximately 35% of the population in developing countries in 1970 to 13% in 2015. In line with the IPCC report mentioned earlier, Dr. Batini also emphasizes that diversity of crop species, crop varieties within species, and genetic diversity within varieties are critical to adapting to climate change. According to her, these genetic resources are the foundation for breeding crops that are resilient to climate change and that can withstand pests and diseases that are likely to become more prevalent with climate change.

On the demand side, Dr. Batini mentions that recently the EAT-Lancet Commission brought together intersectoral experts from health, agriculture, and sustainability to develop targets of healthy diets and sustainable food production. These targets are in line with the Paris Agreement and meet the Sustainable Development Goals (SDGs). The commission concluded that it is possible to provide healthy diets to a growing population by 2050 and beyond through the following five strategies: international and national commitments to healthy diets, reorienting agricultural priorities from quantity of food to quality, intensifying food production to increase high-quality output, strong and coordinated governance related to land and oceans, and halving food losses and waste in line with the SDGs.

Another important aspect is the conservation of land and forest, in fact, the value of healthy soil and intact forests becomes clearer as we approach the long-term 2050 milestone when humanity should transition from net emissions to net sequestration. If we do not keep our soil healthy and our remaining forest intact, it will be impossible to implement the climate goals outlined in the Paris Agreement. But soil cannot continue to sequester carbon if it continues to be disturbed, and intact forests cannot sequester if they are increasingly felled. Falling to account for forgone carbon removals can distort priorities for mitigation action within and beyond the forest sector. Rooting carbon back into the ground through regenerative agriculture is one of the greatest opportunities to address human and climate health. Fiscal policies and structural reforms at both national and international levels are well suited to ensure conservation of land and forests, and successful examples exist of places that have accomplished these goals with benefits for the economy and the environment. Mapping land and accounting for natural capital are essential ingredients for successful conservation efforts, and innovative techniques are now available to do this accurately and globally (Batini, 2021),

Furthermore, to mitigate the adverse effects of climate change on food security, Mahofa (2022) suggests that trade flows are likely to be important in strengthening the resilience of African food systems in the face of shocks emanating from climate change. Delivering food from surplus areas to deficit areas is likely to be important for improving availability and reducing food price volatility from climate shocks, hence the need to improve the movement of food products across borders. Therefore, policies to be adopted may include improving trade facilitation by removing complex regulatory environments and improving transportation infrastructure, reducing intra-SSA tariffs, avoiding trade policy uncertainty and removing export and import bans. The results of his study have implications for improving food security in the face of climate shocks, as the physical availability of food through trade is also an important component of food security. In addition, as climate change will alter specialization patterns, countries are encouraged to concentrate on producing those cereal crops where they have gained comparative advantage as a result of climate change instead of traditional crops they have been used to growing. Given the rich literature in adaptation strategies and policy recommendations, what are the causes of the implementation gaps for both mitigation and adaptation? In the opinion of the 2023 IPCC report, insufficient financing, and a lack of political frameworks and incentives for finance are the main drivers of such gaps (IPCC, 2023). The report adds that annual finance flows targeting adaptation for Africa are billions of USD less than the lowest adaptation cost estimates for near-term climate change. Thus, accelerated financial support for developing countries from developed countries and other sources is a critical enabler to enhance mitigation action.

5.3 Climate Finance

Even though there is no agreed definition of climate finance, the term climate finance, for our purpose, refers to the financial resources devoted to addressing climate change by all public and private actors from global to local scales, including international financial flows to developing countries to assist them in addressing climate change. Climate finance aims to reduce net greenhouse gas emissions and/or to enhance adaptation and increase resilience to the impacts of current and projected climate change. Finance can come from private and public sources, channeled by various intermediaries, and is delivered by a range of instruments, including grants, concessional and non-concessional debt, and internal budget reallocations. While the 2023 IPCC report identifies the following four main types of climate finance: (i) grants, (ii) non-concessional loans (or green bonds), (iii) concessional loans, and (iv) budget reallocation, we will only discuss the first two types of climate finance, as they seem to be more relevant to the Senegalese context.

5.3.1 Grants

Grants provide finance without any repayment requirements. Most grants for adaptation to climate change in Senegal have been provided by multilateral funds such as the Green Climate Fund or a fund managed by a single OECD country such as Germany's International Climate Initiative. It is worth mentioning that some countries have national climate or environment funds that provide grants for their own climate adaptation actions. Grants can also be provided by philanthropic foundations and sometimes by companies as part of their environmental and social responsiveness mandate.

5.3.1.1 Green Climate Fund

The Green Climate Fund (GCF) is the one of the world's largest dedicated funds helping developing countries respond to climate change. It was set up by the United

Nations Framework Convention on Climate Change (UNFCCC) in 2010. The GCF's Initial Resource Mobilization (IRM) in 2014 raised $10.3 billion in pledges, of which $8.2 billion were confirmed through contribution arrangements and after accounting for variations in exchange rates, $7.2 billion have been available for commitment. In December 2019, GCF concluded the first replenishment process with an additional $9.9 billion in pledges (as of August 2020). The Fund seeks to have an impact within eight mitigation and adaptation results areas: agriculture, forestry and other land use; buildings, cities, industries, and appliances; ecosystems and ecosystem services; energy; health, food, and water security; infrastructure; livelihoods of vulnerable communities; and transport. Since its creation, GCF is reported to have funded thirteen projects in Senegal (as reported in Appendix F, Table F.1), with a total financing of $170,1 million.

5.3.1.2 Germany's International Climate Initiative

The International Climate Initiative (IKI) is an important part of the German government's international climate finance commitment. To date, IKI has approved more than 950 climate and biodiversity projects with a total funding volume of almost 6 billion euros (2008–2022). As reported in Appendix F in Tables F.2 and F.3, IKI financed 24 green projects related to climate issues in Senegal.

5.3.1.3 USAID'S Climate Change Program

USAID aims to expand access to clean energy while improving the resilience of Senegal's people and their livelihoods through climate change adaptation measures. Additionally, USAID is integrating a holistic approach in addressing the climate crisis, including its impact on the health of the Senegalese population through expanded malaria interventions, and incorporating climate into school curricula and literacy programming. Since 2013, Power Africa and the Senegal Mission have contributed to the financial close of 803 MW of new generation capacity worth $770M in investments. Power Africa partners have added 302 MWs of new generation capacity contributing to over 58,059 new connections. In 2021, USAID's Power Africa Off-Grid Project achieved 22,706 new connections in Senegal while leveraging $36 million in private sector investment. In 2021, USAID's West Africa Energy Program supported Senegal's clean energy transition by mobilizing $173 million in investment, while facilitating 120 MW in additional generation capacity and 32,726 new connections. A three-year project with Oolu Solar that ended in 2021 brought renewable energy to 8,500 new households, benefiting an estimated 76,500 people and avoiding 484 tons in emissions.

Grants are useful for a range of adaptation actions where it is hard to generate a financial return. These include capacity-building activities, piloting new adaptation innovations, high-risk investment settings or projects where there are considerable non-financial benefits. In contrast, loans and other debt instruments can often support larger investments, for example, for scaling out of successful pilot projects or for building adaptation and resilience into general development investment. To date, a large proportion of international climate finance for adaptation in developing countries, especially in Sub-Saharan Africa, has been grant led, sourced from OECD public funds, indicating that in many instances financing via loans is either considered too risky by the commercial investment sector or it has been hard to demonstrate sufficient return on investment.

5.3.2 *Non-concessional Loans (or Debts)—Green Bonds*

Non-concessional loans (or debts) are commercial instruments, where capital repayment and market interest rates apply. These may be provided through development banks or private banks. Green bonds are a relatively new form of market loan, designed to meet climate and other environmental sustainability criteria in terms of how the proceeds are used. In recent years, green bonds have offered better interest than ordinary bonds owing to oversubscription by investors who are looking to move toward environmentally sustainable investment portfolios. Indeed, green bonds are among a blend of financial instruments that could be catalytic in facilitating this transition in a just manner through responsible, sustainable investments.

Green bonds are a useful tool for mobilizing additional capital to finance sustainability challenges in Africa but the uptake there has been low for a number of reasons. It is both promising and encouraging that various existing frameworks have made it possible for some African states to issue green bonds in recent years. Nonetheless, the 2020 Stockholm Sustainable Finance Centre report identified the following obstacles to increasing the pace of development of the African green bond market: (i) frameworks, (ii) capacities, (iii) supporting regulation, (iv) lack of verifiers, (v) lack of incentive structures (Marbuah, 2020).

However, the report highlights that Africa presents many possibilities for investors interested in climate-smart agriculture, renewable energy, energy efficiency, and waste management, among other things. Several sectors on the continent are in need of significant green investments. Major sustainability concerns require urgent financial solutions. Green bonds could make a difference in either reversing or avoiding future environmental catastrophes. Senegal has signed and ratified the Paris Agreement, which entered into force on November 4th, 2016. These commitments to a greener, climate aligned, and resilient economy are suggestive of governments' inclinations to foster low-carbon transitions.

5.4 Financialization of Crops (International Commodity Markets)

Despite agriculture's importance, Senegal, as we have shown earlier, lies within the drought-prone Sahel region, with irregular rainfall and generally poor soils. As a result, Senegal relies on imports to meet approximately 70% of its food needs. The top food and beverage suppliers to Senegal are the European Union (48%), India (26%), Brazil (10%), China (10%), and Indonesia (6%). Imports from the United States account for less than 1% of Senegal's total food imports. Top agricultural imports in Senegal are rice, wheat, corn, onions, palm oil, sugar, and potatoes. The top imports from the United States are feed grains (corn) and consumer-oriented products which include: sauces, mixed condiments, seasonings, fish oil, spices, and non-alcoholic beverages.[4] Business within Senegalese agricultural sectors is based on cash, which creates several challenges for producers: (i) lack of security as many cases of theft or loss are frequently experienced after cash payments, (ii) difficult financial management due to the fact that farmers receive income only during harvest periods and often lack resources the rest of the year, and (iii) dependence of farmers on cash payments, which are highly volatile. This dependence limits their ability to access other financial services such as credit, savings, or even insurance.[5]

5.4.1 Booming Groundnuts Export to China

Senegal is one of Africa's top peanut-exporting countries. Exports to China yield significant benefits for local farmers. But with them come long-term repercussions for local consumption, processing industries, and future production. Indeed, Senegal is among the top exporting countries of peanuts globally. In 2021, Senegal ranked as Africa's second exporter with an overall export value of 285 million USD, amounting to 6.85% of global peanut exports—a figure that was only topped by Sudan. In 2021, peanuts were the fifth most-exported product in Senegal, amounting to 5.4% of Senegal's total exports. China alone imported more than 323,000 tons of peanuts from Senegal in 2020, equivalent to more than 90% of Senegal's total peanut exports. In light of these volumes, it is not surprising that the cultivation and export of peanuts offer considerable economic opportunities for farmers, traders, and the state. Peanuts are considered the backbone of the national economy, with the majority of rural households depending on groundnuts for their livelihoods as well as for subsistence crop production, export, processing, and income generation.[6] At first sight, large exports to China appear to be a good opportunity for Senegal to improve its account

[4] USDA, Foreign Agricultural Service, Senegal: Exporter Guide, December 30,2021.

[5] Consultative Group to Assist the Poor, CGAP Annual Report 2016: Advancing Financial Inclusion to Improve the Lives of the Poor (Washington DC, July 2016).

[6] Senegal Groundnut Value Chain Competitiveness and Prospects for Development (2017). https://doi.org/10.1596/28399.

balance and to contribute to domestic growth and employment. However, this strategy comes with several major caveats. Strong external demand for raw nuts and the resulting higher prices penalize local processing industries specifically. As a result, local and international buyers, namely Chinese, compete over the produce—with the Chinese systematically outpricing local buyers. Processing plants (mainly small and medium-sized firms) are left with limited quantities of peanuts to turn into products such as oil and peanut butter. Production capacities that could contribute to local value-creation remain unused, hindering growth and job-creation, and as such replicating a situation that had led to export restrictions in the past. In addition to natural limits that cap the overall amount of available peanuts, several challenges add to the unsustainability of current economic practices in the industry: (1) the limited use of modern irrigation technologies, and thus farmers' over-reliance on rainfalls to ensure food production, (2) the effects of climate change (i.e., with increasing droughts and floods), and (3) the growing scarcity of arable land worsened by rapid urbanization.

5.4.2 Cereal Markets

Senegal's experience with cereals market regulation grows out of a colonial legacy that accepts a preeminent role for the state as protector of producers and consumers. Yet, the reality of cereals market operation in Senegal differs substantially from that specified in regulatory texts, and satisfaction of certain performance objectives is achieved largely because of illegal parallel market activity involving violations of official rules and policies (Morris & Newman, 1989). These two researchers added that cereals traders in particular continue to be maligned, frequently being accused of buying grain from farmers at extremely low prices, hoarding stocks in order to create artificial shortages, selling to consumers at excessively high prices, and charging usurious rates of interest. These common perceptions produce an unfavorable impression of cereal marketing as an economic activity and contribute to a strong bias against the private sector intermediaries or "middlemen" who perform grain trading operations. To them, contrary to the conventional wisdom, intermediaries in parallel (or unconventional) cereals markets frequently are able to provide a wider range of marketing services at lower cost than the state and parastatal marketing organizations, which are intended to supplant them.

5.4.2.1 Organization of Official Cereals Markets

Rice Whether produced locally or imported, rice is distributed through a network of two licensed private traders. First, wholesalers are defined as traders who sell lots of grain greater than 200 kilograms. They require a special wholesaler's license issued by the Ministère du Commerce Intérieur (MCI), additionally the trader must have access through ownership or rental to a grain storage facility which conforms to government health and safety regulations. Licensed wholesalers are authorized to purchase rice at official wholesale prices from Caisse de Péréquation et Stabilization des Prix (CPSP) or from one of four authorized private importers. Wholesalers may then transport rice to regional consumption centers for sale to retailers. Official transport rates for cereals are established by the government. Interregional transport of rice requires a transfer permit, which can be obtained from the MCI for no charge. Second, retailers, who are defined as traders who sell lots of grain less than 200 kg, require a general retailer's license. These are issued by the MCI for a token fee and confer the right to sell rice and other cereals, as well as a wide range of staple foods and dry goods. Licensed retailers are authorized to purchase grain from wholesalers at official wholesale-retail prices and to sell to consumers at official retail prices, both of which vary by region to accommodate differential transport rates paid by wholesalers. Rice and other cereals are also sold through a chain of retail stores operated by the Société Nationale de Distribution (SONADIS), the parastatal charged with distributing consumer goods throughout the country. An important objective of SONADIS participation in the market is to ensure the ready availability of rice, millet, sorghum, maize, and other staples at official retail prices. During periods when supplies are available from SONADIS, the presence of these retail stores places an upper limit on the prices that private retailers can effectively charge.

Coarse grains Coarse grains are millet, sorghum, and maize. Unlike the case of rice, the state no longer attempts to exercise a legal monopoly over the assembly and processing of coarse grains; these functions are now shared with private traders, who operate according to a set of official regulations and policies. The government specifies who may participate in what marketing functions, when and where transactions may take place, and what prices may be charged. The official marketing channel for coarse grains begins with a system of assemblers who purchase small quantities at the village and periodic market level. These assemblers include traders working for their own account, commission agents for wholesalers, and others buying grain for personal consumption. All traders assembling grain are required to possess at least a retailer's license, and those assembling quantities greater than 200 kg are required to possess a wholesaler's license. The government announces the date at which assembly operations may begin (known as the opening of the annual marketing campaign) and specifies an official producer price for each type of grain. Licensed private traders are authorized to transport coarse grains to regional consumption centers for sale, storage, or consumption. Official transport rates are established by the government. Transport of coarse grains outside the production region is prohibited without a transfer permit, which can be obtained from the MCI for no charge. A government agency, CSA, also enters the market to purchase coarse grains, particularly millet, either by direct purchases or through distribution of funds to finance assembly by cooperatives. The stated purposes of CSA participation in the market are both to support the official producer prices and to enable the government to build up grain reserves which can be used later in the year during the hungry season to maintain adequate supplies for urban consumers.

5.4.2.2 The Role of the Parallel Markets

Rice Dissatisfaction with the official marketing system has led to the emergence of a thriving parallel market for rice in the Senegal River valley. Seasonal itinerant traders are the main actors in this market. Traveling throughout the production zones during the months immediately following the harvest, they purchase paddy directly from farmers, transport it by animal cart or bush taxi to nearby village hullers for processing, sell the husks-and-bran byproduct to local herders for use as animal feed, and send the processed rice by taxi or truck to urban consumption centers in the south (Morris & Newman, 1989). The empirical evidence of Morris and Newman fails to substantiate the view espoused by many government officials and policymakers that parallel market rice traders in the Senegal River valley are inefficient and exploitative. On the contrary, their survey data indicate that the parallel market trade in many respects compares favorably with official marketing operations. According to farmers, the itinerant traders respond quickly and effectively to changes in market conditions, rapidly identifying regions with surplus paddy, taking immediate delivery of grain, and offering on-the-spot cash.

Coarse grains Even though private traders offer higher prices than the official producer price, they generally are able to achieve satisfactory net marketing margins by reducing marketing costs well below those incurred by CSA (Morris & Newman, 1989). Thus, as in the case of rice, the failure of private traders to comply with the official pricing structure does not mean that they earn excessive profits on their marketing activities. Rather, by ignoring the rigid official price structure, traders are able to adjust offer and sale prices to take advantage of current market opportunities, buying when and where surplus grain is available and selling when and where deficits exist. This would be difficult, if not impossible, were they to adhere strictly to the official price structure.

Ultimately, Senegal's heterogeneous cereals marketing system, like those in many francophone African countries, includes public marketing organizations, parastatals, authorized private traders, and parallel market intermediaries. What is noteworthy about the Senegalese system is the important role played by parallel market participants. While the theory of parallel markets does not predict economic inefficiency, the policy dialogue in many countries continues to be driven by the belief that parallel markets unnecessarily disadvantage producers and consumers through inflated marketing costs and unfavorable prices. In northern Senegal, the parallel market serves as the main marketing channel for locally produced rice. Rice producers in the Senegal River valley who sell to illegal, itinerant traders find a reliable and accessible outlet for their production at higher than official prices, which encourages increased production and marketing. In central Senegal, the parallel market is equally important in facilitating the marketing of coarse grains. Coarse grains producers throughout the Groundnut Basin find a ready outlet for their production that might not be available if private traders were not willing to engage in illegal marketing activities. By ignoring official prices and marketing margins, traders are able to transport grain to deficit regions and to store grain into the hungry season, with the result that consumers have larger supplies and face lower prices than might otherwise have prevailed. The empirical study of Morris and Newman underscores the importance of applied research on parallel markets as input into the food policy dialogue. Agricultural policy makers in Senegal continue to operate without the benefit of adequate empirical information. This lack of knowledge often leads to the design and implementation of policies that are not only misguided, but actually counterproductive in the sense that they frequently impede rather than facilitate the activities of private traders. Thus there exists a continuing need for empirical research of immediate relevance to policymakers.

References

Batini, N. (Ed.). (2021). *The economics of sustainable food: Smart policies for health and the planet*. Island Press.

Hathie, I., Seydi, B., Samaké, L., & Sakho-Jimbira, S. (2017). Reducing economic and environmental vulnerability to achieve food and nutrition security in Senegal. Brookings Institution. [134]

IPCC. (2023). Climate Change 2023: Synthesis Report. Contribution of Working Groups I, II and III to the Sixth Assessment Report of the Intergovernmental Panel on Climate Change [Core Writing Team, H. Lee and J. Romero (eds.)] (pp. 35–115). IPCC. https://doi.org/10.59327/IPCC/AR6-9789291691647 [135]

Mahofa G. (2022). Climate Change and Agricultural Trade in Sub-Saharan Africa. Research Paper 499, African Economic Research Consortium. Retrieved May 12, 2023, from http://publication.aercafricalibrary.org/ [123]

Marbuah, G. (2020). Scoping the sustainable finance landscape in Africa. Stockholm Sustainable Finance Centre. https://www.stockholmsustainablefinance.com/scoping-the-sustainable-finance-landscape-in-africa-ssfc-reprt/ [136]

Morris, M. L., & Newman, M. D. (1989). Official and parallel cereals markets in Senegal: Empirical evidence. *World Development, 17*(12), 1895–1906. https://doi.org/10.1016/0305-750X(89)90269-6 [137]

Chapter 6
Conclusion

This dissertation was an effort to estimate the resulting impact of climate change on long-term crop production, food security, and child malnutrition. Chapter 1 informed us about the historical climate trends with the visualization of annual maximum and minimum temperatures as well as average annual rainfall from 1900 to 2020. This data showed that the country, especially after the severe drought event of the 1970s, has warmed up so much that there is no turning back. Chapters 2 and 3 forecasted the long term cereal production of the four major staple crops in the country, which came out to be very vulnerable to not only the different scenarios of greenhouse gas emissions and planting areas, but also to changes in technology and the quality of soil. These have serious implications to children's malnutrition, as highlighted in Chap. 4. Climate variability and change stand to exacerbate malnutrition through effects on both the natural and social systems upon which humans depend, including food systems. The last chapter of this study discussed financialization by pointing out the importance of the parallel markets as well as the recent rush of overseas investors to Senegalese land.

While the emissions of the West are responsible for the climate crisis and its consequences, Africa now has to pay for the climate damage and Senegal is no exception to this. Our study showed that without adaptation strategies, the number of malnourished children significantly increased as the amount of available food could not meet the needs of the growing population. Based on the rich literature of climate change and agricultural production, the biggest challenge facing emerging economies and developing countries to transition to net zero is climate finance. The sub-Saharan African region faces a huge shortfall of infrastructure investment relative to that transition to net zero needs. Indeed, trillions of US dollars are needed, but we are talking about millions of US dollars when it comes to climate change adaptation in agricultural production. On one hand, international institutions are estimated at $100 billion to be a combination of public and private funds. While, on the other hand, G20 estimates $1 trillion with $500 billion have to come from the

M. Toure, *Agricultural Adaptation to Climate Change in Sub-Saharan Africa*,
Advances in African Economic, Social and Political Development,
https://doi.org/10.1007/978-3-031-80716-9_6

private sector in the form of equity and debt. The other $500 billion have to come from public finance funding where $350 billion comes from multilateral banks. When investigating public expenditure on agriculture in Africa, we found that in many African countries, public expenditure on agriculture is low, and policy is anti-agriculture. Also, private investment in agriculture is not forthcoming in countries, which accord it low priority, and which have poor policy environments for agriculture. Furthermore, subsidies for irrigation water, chemical fertilizers induce overuse of these inputs, rather than water saving and use of organic materials for fertilization. Moreover, African agriculture research, extension, and education institutions are often neglected. They are the source of new local knowledge of climate impact and adaptation.

Though children are the least responsible for climate change, they will bear the greatest burden of its impact. From droughts to floods to wildfires to deforestation, climate change and environmental degradation is a direct threat to a child's ability to survive, grow, learn, and thrive. The world is suffering a triple burden of child malnutrition, combining undernutrition, micronutrient deficiencies, and overweight and obesity. Based on UNICEF's database, as referred in Chap. 4, 144 million children under 5 are stunted, 47 million children under 5 are wasted, 340 million children under 5 are micronutrient deficient, and 38 million children under 5 are overweight. These challenges are only exacerbated by the compounding effects of climate change and environmental degradation. Climate variability and extremes are eroding and reversing progress in ending child hunger and malnutrition. As shown in Chaps. 2 and 3, increases in temperature, heat waves, droughts, and floods result in declines in crop yields, thus affecting food quality, diversity, nutrient density, safety, and prices. This in turn affects all dimensions of food security and nutrition, including food availability, access, utilization, and stability. The Intergovernmental Panel on Climate Change predicts that the biggest threat to health from climate change will be through the effects on nutrition outcomes. In addition, environmental degradation threatens food safety and nutrition through both the loss of biodiversity and the contamination of water, air, soil, and exposure to antibiotics and pesticides. Specifically, biodiversity supports critical dietary diversity, which is essential for food security and nutrition, but is increasingly difficult to guarantee due to the rise of homogenous diets where rice, wheat, and maize represent 2/3rds of global caloric intake. Such homogenous diets limit children's dietary intake of micronutrients while increasing the exposure to calorie dense foods prone to overweight and obesity.

One potential solution might involve matching private finance and green infrastructure. In that case, the literature points out that the biggest challenge is to build bridges between, on the one hand, infrastructure investment needs in non-advanced countries and, on the other, private sources of finance abundant in dollars and other convertible currencies with few opportunities to obtain returns compatible with their requirements on their liability side. In practice, building such bridges requires the completion of two tasks. First, the development of properly structured projects, with risks and returns in line with the preferences of the different types of financial intermediation, would help close the private financing gap in infrastructure. The second task to boost private infrastructure investment in emerging and developing economies is the reduction of legal, regulatory, and political risks. Transparency and harmonization of rules and standards can increase the scale of comparable projects and make it possible to build project portfolios. Non-banking financial institutions often highlight the absence of large enough project portfolios as a deterring the setting up of business lines focused on the area. This is a particular weakness in the case of smaller countries.

Another interesting adaptation strategy involves the creation of a green bank. The proposed Green Bank would be different from existing multilateral development banks: (1) it would include private shareholders as well as governments; (2) voting rights would be organized so that countries of the Global South would have the same voice as countries of the Global North and private shareholders; and (3) it would only finance green projects which could be national, regional, or global. Ultimately, concrete actions need to be done to achieve a green transition in sub-Saharan African countries, there is the need to act now by combining the story of growth and climate responsibility. Otherwise, it is closed to climate denial or to assume that climate change projection is not that bad.

Appendix A
FAO Supplementary Data

A.1 Senegal's Detailed Soil Characteristics

See Fig. A.1.

Reference

Han, E., Ines, A. V., & Koo, J. (2019). Development of a 10 km resolution global soil profile dataset for crop modeling applications. *Environmental Modelling & Software, 119*, 70–83. [40]

A.2 Genetic Coefficients of Rice, Maize, Sorghum, and Pearl Millet Cultivars Used in this Study

See Tables A.1, A.2, A.3, and A.4.

References

de Vries, M. E., Sow, A., Bado, V. B., & Sakane, N. (2012). Simulation of potential yields of new rice varieties in the Senegal River Valley. In *Improving soil fertility recommendations in Africa using the decision support system for agrotechnology transfer (DSSAT)* (pp. 141–155). Springer, Dordrecht. [43]

Singh, P., Nedumaran, S., Traore, P. C. S., Boote, K. J., Rattunde, H. F. W., Prasad, P. V. V., Singh, N. P., Srinivas, K., & Bantilan, M. C. S. (2014). Quantifying potential benefits of drought and heat tolerance in rainy season sorghum for adapting to climate change. *Agricultural and Forest Meteorology, 185*, 37–48. [45]

Singh, P., Boote, K.J., Kadiyala, M. D. M., Nedumaran, S., Gupta, S. K., Srinivas, K., & Bantilan, M. C. S. (2017). An assessment of yield gains under climate change due to genetic modification of pearl millet. *Science of the Total Environment, 601*, 1226–1237. [46]

M. Toure, *Agricultural Adaptation to Climate Change in Sub-Saharan Africa*, Advances in African Economic, Social and Political Development,
https://doi.org/10.1007/978-3-031-80716-9

```
*SN03855414     SEN    LoamySand    200     ISRIC soilgrids + HC27
@SITE         COUNTRY            LAT      LONG SCS Family
 -99              SN        15.625  -15.458      HC_GEN0021
@ SCOM  SALB  SLU1  SLDR  SLRO  SLNF  SLPF  SMHB  SMPX  SMKE
    BK  0.15  4.00  0.75 65.00  1.00  0.60 SA001 SA001 SA001
@  SLB  SLMH  SLLL  SDUL  SSAT  SRGF  SSKS  SBDM  SLOC  SLCL  SLSI  SLCF  SLNI  SLHW  SLHB  SCEC  SADC  SLPX
     5 A     0.046 0.094 0.397  1.00  6.77  1.32  1.92  8.60  6.14 -99.0  0.12  6.55 -99.0  11.1 -99.0  10.4
    15 A     0.053 0.102 0.394  0.90  5.96  1.34  1.63  9.77  5.97 -99.0  0.09  6.62 -99.0   9.7 -99.0  10.3
    30 AB    0.065 0.115 0.390  0.80  4.92  1.37  1.24 11.56  5.57 -99.0  0.07  6.72 -99.0   9.4 -99.0  10.1
    60 BA    0.074 0.126 0.388  0.60  4.20  1.42  0.80 13.03  5.23 -99.0  0.06  6.83 -99.0   9.8 -99.0   8.0
   100 B     0.073 0.124 0.388  0.00  4.29  1.48  0.46 12.91  4.90 -99.0  0.00  6.97 -99.0   9.9 -99.0   8.0
   200 BC    0.064 0.113 0.390  0.00  4.99  1.54  0.26 11.57  4.73 -99.0  0.00  7.15 -99.0   9.9 -99.0   8.0

*SN03950477     SEN    SandClayL    200     ISRIC soilgrids + HC27
@SITE         COUNTRY            LAT      LONG SCS Family
 -99              SN        13.792  -13.542      HC_GEN0011
@ SCOM  SALB  SLU1  SLDR  SLRO  SLNF  SLPF  SMHB  SMPX  SMKE
    BK  0.10  6.00  0.50 75.00  1.00  0.60 SA001 SA001 SA001
@  SLB  SLMH  SLLL  SDUL  SSAT  SRGF  SSKS  SBDM  SLOC  SLCL  SLSI  SLCF  SLNI  SLHW  SLHB  SCEC  SADC  SLPX
     5 A     0.160 0.266 0.400  1.00  0,62  1.34  3.64 26.53 22.22 -99.0  0.12  6.60 -99.0  29.7 -99.0  10.4
    15 A     0.171 0.279 0.403  0.85  0.50  1.36  3.09 28.43 21.42 -99.0  0.09  6.67 -99.0  26.0 -99.0  10.3
    30 AB    0.187 0.297 0.409  0.70  0.37  1.39  2.35 31.12 20.50 -99.0  0.07  6.77 -99.0  25.2 -99.0  10.1
    60 BA    0.202 0.314 0.414  0.50  0.27  1.44  1.51 33.69 19.48 -99.0  0.06  6.87 -99.0  26.3 -99.0   8.0
   100 B     0.201 0.312 0.413  0.38  0.27  1.50  0.88 33.55 18.77 -99.0  0.05  7.02 -99.0  26.5 -99.0   8.0
   200 BC    0.191 0.299 0.408  0.05  0.34  1.55  0.50 31.87 18.23 -99.0  0.05  7.19 -99.0  26.4 -99.0   8.0

*SN03989327     SEN    SandClayL    200     ISRIC soilgrids + HC27
@SITE         COUNTRY            LAT      LONG SCS Family
 -99              SN        13.042  -16.042      HC_GEN0011
@ SCOM  SALB  SLU1  SLDR  SLRO  SLNF  SLPF  SMHB  SMPX  SMKE
    BK  0.10  6.00  0.50 75.00  1.00  0.60 SA001 SA001 SA001
@  SLB  SLMH  SLLL  SDUL  SSAT  SRGF  SSKS  SBDM  SLOC  SLCL  SLSI  SLCF  SLNI  SLHW  SLHB  SCEC  SADC  SLPX
     5 A     0.153 0.264 0.399  1.00  0.64  1.30  2.65 25.33 25.90 -99.0  0.12  5.58 -99.0  19.1 -99.0  10.4
    15 A     0.163 0.276 0.402  0.85  0.52  1.32  2.24 27.17 25.22 -99.0  0.09  5.66 -99.0  16.7 -99.0  10.3
    30 AB    0.179 0.293 0.408  0.70  0.39  1.35  1.71 29.82 23.96 -99.0  0.07  5.75 -99.0  16.2 -99.0  10.1
    60 BA    0.194 0.310 0.413  0.50  0.29  1.40  1.10 32.41 22.63 -99.0  0.06  5.86 -99.0  17.0 -99.0   8.0
   100 B     0.194 0.308 0.412  0.38  0.29  1.46  0.64 32.28 22.07 -99.0  0.05  6.00 -99.0  17.1 -99.0   8.0
   200 BC    0.183 0.295 0.407  0.05  0.36  1.52  0.38 30.57 21.82 -99.0  0.05  6.18 -99.0  17.1 -99.0   8.0
```

Fig. A.1 Soil characteristics location in Senegal used for DSSAT simulation. *Source* "Development of a 10 km resolution global soil profile dataset for crop modeling applications" by Han et al. (2019)

Table A.1 Genetic coefficients of rice cultivar used in the study

Code*	Cultivar (Modified IR 64)	Description
P1	872	Time period from seedling emergence (expressed as growing degree days [GDD] in °C-d above base temperature of 9 °C)
P20	12	Critical photoperiod (in hours) at which the development occurs at a maximum rate
P2R	160	Extent to which phasic development leading to panicle initiation is delayed (expressed as GDD in °C-d) for each hour increase in photoperiod above P20
P5	600	Time period in GDD °C-d) from beginning of grain filling to physiological maturity with a base temperature of 9 °C
G1	60	Potential spikelet number coefficient as estimated from the number of spikelets per g of main culm dry weight at anthesis
G2	0.025	Single grain weight (g) under ideal growing conditions
G3	1	Tillering coefficient relative to IR64 cultivar under ideal conditions
G4	1	Temperature tolerance coefficient
PHINT	83	Time interval in degree-days for each leaf-tip to appear under non-stressed conditions

Source "Simulation of potential yields of new rice varieties in the Senegal River Valley. In: Improving Soil Fertility Recommendations in Africa using the Decision Support System for Agrotechnology Transfer (DSSAT)" by de Vries et al. (2012)

Table A.2 Genetic coefficients of maize cultivar used in the study

Code	Cultivar (Massongo)	Description
P1	260	Thermal time from seedling emergence to the end of the juvenile phase (expressed in degree days above a base temperature of 8 °C)
P2	0.75	Extent to which development (expressed as days) is delayed for each hour increase in photoperiod above the longest photoperiod at which development proceeds at a maximum rate
P5	800	Thermal time from silking to physiological maturity (expressed in GDD above a base temperature of 8 °C)
G2	750	Maximum possible number of kernels per plant
G3	5.3	Kernel filling rate during the linear grain filling stage and under optimum conditions (mg day^{-1}).
PHINT	49	The interval in thermal time (degree days) between successive leaf tip appearances

Source "Intensifying maize production under climate change scenarios in central West Burkina Faso" by Worou et al. (2018)

Worou, O. N., Tondoh, J. E., Sanou, J., Gaiser, T., Nikiema, P. M., Bayala, J., Bazié, P., Ky-Dembele, C., & Kalinganiré, A. (2018). Intensifying maize production under climate change scenarios in central West Burkina Faso. In W. L. Filho (Ed.), *Handbook of climate change resilience* (pp. 1–23). Springer International Publishing.

Table A.3 Genetic coefficients of sorghum cultivar used in the study

Code*	Cultivar (CSM 335)	Description
P1	413	Thermal time from seedling emergence to the end of the juvenile phase
P2	102	Thermal time from the end of the juvenile stage to tassel initiation under short days
P20	12.8	Critical photoperiod or the longest day length (in hours) at which development occurs at a maximum rate
P2R	280	Extent to which phasic development leading to panicle initiation (expressed in degree days) is delayed for each hour increase in photoperiod above P20
PANTH	617.5	Thermal time from the end of tassel initiation to anthesis (degree days).
P3	152.5	Thermal time from the end of flag leaf expansion to anthesis (degree days)
P4	81.5	Thermal time from anthesis to beginning grain filling (degree days)
P5	640	Thermal time from beginning of grain filling to physiological maturity (degree days)
PHINT	49	The interval in thermal time between successive leaf tip appearances (degree days)
G1	3	Scaler for relative leaf size
G2	4	Scaler for partitioning of assimilates to the panicle

Source "Quantifying potential benefits of drought and heat tolerance in rainy season sorghum for adapting to climate change" by Singh et al. (2014)

Table A.4 Genetic coefficients of pearl millet cultivar used in the study

Code*	Cultivar (CIVT)	Description
P1	365	Thermal time from seedling emergence to the end of the juvenile phase (expressed in degree days above a base temperature of 10 °C)
P20	12	Critical photoperiod or the longest day length (in hours) at which development occurs at a maximum rate
P2R	260	Extent to which phasic development leading to panicle initiation (expressed in degree days) is delayed for each hour increase in photoperiod above P20
P5	285	Thermal time (degree days above a base temperature of 10 °C) from beginning of grain filling to physiological maturity
G1	0.6	Scaler for relative leaf size
G4	0.73	Scaler for partitioning of assimilates to the panicle
PHINT	43	The interval in thermal time (degree days) between successive leaf tip appearances
GT	1.2	Tillering coefficient, equivalent to G1, but on tillers
G5	11	Potential grain size (mg)

Source "An assessment of yield gains under climate change due to genetic modification of pearl millet" by Singh et al. (2017)

A.3 Major Cereal Crops in Senegal for 1989–2020

See Table A.5.

Table A.5 Area, production, and yield of major cereal crops in Senegal for 1989–2020

	Rice			Maize			Sorghum		
Year	Area (ha)	Yield (kg/ha)	Production (tons)	Area (ha)	Yield (kg/ha)	Production (tonnes)	Area (ha)	Yield (kg/ha)	Production (tonnes)
1989	79331	2120	168227	92990	1413	131407	131461	965	126868
1990	72957	2483	181119	116686	1141	133147	172860	910	157323
1991	72264	2350	169800	90632	1132	102633	99636	784	78094
1992	73696	2407	177346	105041	1091	114564	131398	888	116739
1993	77945	2481	193374	108960	1269	138318	126360	784	99073
1994	77736	2087	162228	106814	1013	108233	141947	867	123092
1995	68966	2250	155152	97889	1088	106509	148414	858	127328
1996	73811	2016	148780	84913	1044	88634	148646	895	133009
1997	74698	2325	173702	62178	1291	80281	154476	766	118297
1998	45405	2720	123519	53714	826	44339	201756	593	119574
1999	130000	2800	364000	70440	939	66132	230196	641	147444
2000	86252	2345	202293	70715	1111	78593	165394	869	143750
2001	87230	2373	206989	88399	1228	108546	174724	857	149649
2002	76025	2268	172395	108114	743	80372	199757	585	116929
2003	87814	2640	231805	175575	2283	400907	208363	911	189787
2004	81486	2476	201744	147298	2719	400555	163130	775	126492
2005	97779	2854	279080	143039	2796	399958	149173	965	143989
2006	85037	2240	190493	130461	1392	181585	159063	761	121003
2007	80312	2408	193379	143769	1101	158266	155919	646	100704
2008	125329	3257	408219	216517	1835	397326	249297	1009	251515
2009	139388	3602	502104	178710	1630	291301	240425	936	224956
2010	147208	4103	604043	121235	1538	186511	174264	933	162599
2011	109177	3717	405824	109517	1133	124092	135960	639	86865
2012	117729	3989	469649	147699	1547	228423	143271	960	137574
2013	108547	4018	436153	156576	1443	225902	108803	846	92029
2014	134973	4142	559021	146361	1221	178732	125066	818	102323
2015	237300	3819	906348	199502	1525	304296	200737	939	188500
2016	260854	3627	946000	148467	1471	218450	126912	997	126515
2017	305934	3306	1011269	231659	1771	410364	221329	974	215491
2018	323635	3728	1206587	250891	1936	485673	247678	1193	295463
2019	345596	3444	1155730	258391	2054	530704	238833	1131	270168
2020	407527	3343	1349723	286977	2655	761883	278101	1357	377322

Source FAOSTAT database (2022)

Reference

FAOSTAT database. Retrieved May 15, 2020 from http://www.fao.org/faostat/en/#data [32]

Appendix B
Predicted Climate Data for 2050 and 2080

B.1 Picked Climate Scenarios

MarkSim is a weather generator that uses 720 classes of weather, worldwide, to calculate the coefficients of a third-order Markov rainfall generator. This constitutes 'stochastic downscaling' as it fits a Markov model to the GCM output and uses it to generate weather data for the site indicated. The third weapon in MarkSimGCM is built into the weather typing in the clustering process. The 720 classes of world weather define each set of regression equations that MarkSim uses to determine the coefficients for the modeling process. When a climate changes, such that it no longer applies to the original class, then the whole regression structure changes. This means that a changing climate will be modeled by the one most like it in the real world.

The only drawback is that the system cannot model completely new climates except by extrapolation of the regression models from the nearest cluster. But then we are still to see a new climate in sufficient detail to fit the model to it and GCM results are not precise enough to do this for the future. In the meantime, we will have to wait. Figure A.2.1 shows the GCMs used for this study due to their thinny latitude and longitude resolution (Table B.1).

B.2 Representative Concentration Pathways (RCPs)

The findings of 2013–14 Intergovernmental Panel on Climate Change (IPCC) Fifth Assessment Report (AR5) were based on a new set of scenarios that replace the Special Report on Emissions Scenarios (SRES) standards employed in two previous reports. The new scenarios are called Representative Concentration Pathways (RCPs). The RCPs describe 4 different scenarios based on different assumptions about population, economic growth, energy consumption and sources and land use

M. Toure, *Agricultural Adaptation to Climate Change in Sub-Saharan Africa*, Advances in African Economic, Social and Political Development,
https://doi.org/10.1007/978-3-031-80716-9

Table B.1 Three out of 17 picked general circulation models (GCMs) available in MarkSimGCM

Model	Institution	Resolution Lat × Long!	References
CSIRO-Mk3.6.0	Commonwealth Scientific and Industrial Research Organisation and the Queensland Climate Change Centre of Excellence	1.875 × 1.875	Collier et al. (2011). The CSIROMk3.6.0 Atmosphere-Ocean GCM: participation in CMIP5 and data publication. MODSIM (2011), Perth, 12–16 December 2011
HadGEM2-ES	Met Office Hadley Centre	1.2414 × 1.875	Collins et al. (2011). Development and evaluation of an Earth-System model-HadGEM2. GMD 4(4):1051–1075
MIROC5	Japan Agency for Marine-Earth Science and Technology, Atmosphere and Ocean Research Institute (The University of Tokyo), and National Institute for Environmental Studies	1.4063 × 1.4063	Watanabe et al. (2010). Improved Climate Simulation by MIROC5: Mean States, Variability, and Climate Sensitivity. J. Climate, 23, 6312–6335

Source "Very High Resolution Interpolated Climate Surfaces of Global Land Areas. International Journal of Climatology" by Hijmans et al. (2005)

over this century. There are four RCPs (Representative Concentration Pathway) scenarios which are based on multi-gas emission scenarios, namely RCP2.6, RCP4.5, RCP6, and RCP8.5, are labeled after a possible range of radiative forcing values in the year 2100 relative to pre-industrial values: +2.6, +4.5, +6.0, and +8.5 W/m^2, respectively and as shown in Fig. A.2.2. For each category of emissions, an RCP contains a set of starting values and the estimated emissions up to 2100, based on assumptions about economic activity, energy sources, population growth, and other socio-economic factors. They will also facilitate and expedite future climate change assessments across the integrated assessment community.

The RCP 2.6 is developed by the IMAGE modeling team of the Netherlands Environmental Assessment Agency. The emission pathway is representative for scenarios in the literature leading to very low greenhouse gas concentration levels. The RCP2.6 is known as "Low emissions'. The RCP2.6 is not comparable with any SRES scenario. This future would require:Low energy intensity, CO_2 emissions stay at today's level until 2020, then decline and become negative in 2100, A world population of 9 billion by year 2100, Declining use of oil, and Methane emissions reduced by 40%.

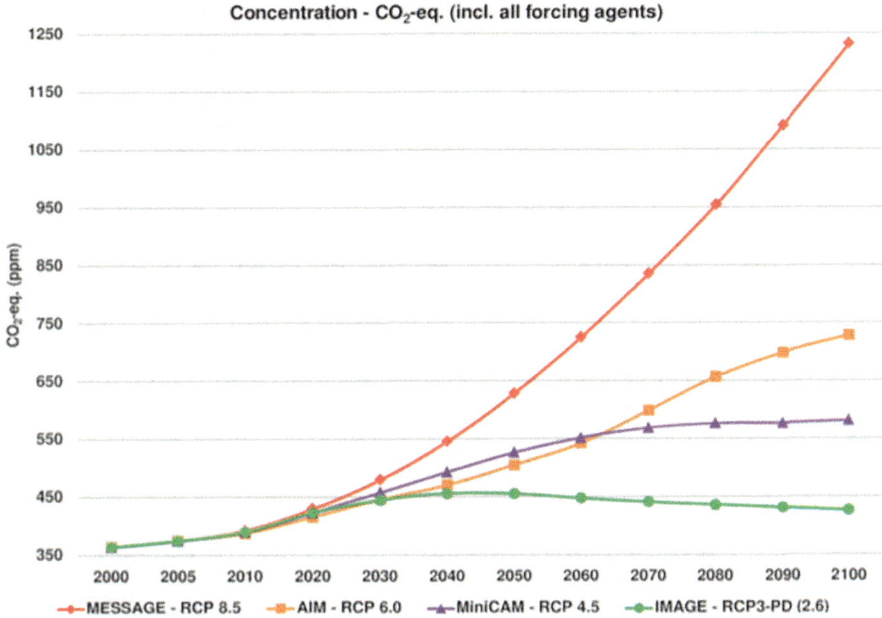

Fig. B.1 Representative concentration pathways. *Source* "Very High Resolution Interpolated Climate Surfaces of Global Land Areas. International Journal of Climatology" by Hijmans et al. (2005)

RCP8.5's radiative forcing levels by the end of 2100 are around 8.5 W/m^2 under our 'best-estimate' set of model parameters with forcing levels increasing further thereafter—up to 12 W/m^2 by 2250, when concentrations stabilize. This RCP is consistent with a future with no policy changes to reduce emissions. The RCP8.5 is known as 'High emissions'. The RCP8.5 is comparable with SRES scenario A1F1. This future is consistent with: Rapid increase in methane emissions, Heavy reliance on fossil fuels, High energy intensity, A world population of 12 billion by 2100, and Increased use of croplands and grassland which is driven by an increase in population (Fig. B.1).

Reference

Hijmans, R., et al. (2005). Very high resolution interpolated climate surfaces of global land areas. *International Journal of Climatology, 25*, 1965–1978. https://doi.org/10.1002/joc.1276

[illegible]

[illegible]

[illegible]

Appendix C
About the Crop Model—DSSAT

C.1 Variables and Parameters

See Table C.1.

C.2 Equations

The DSSAT crop model is an open source software, which operates in Windows, Linux, and iOS and FORTRAN is its programming language. The following sets of equations not only reproduce the response of the DSSAT crop model, but it also reproduces the response of global gridded crop models [60]. For any given day d of the growing season (defined here as the period between the planting day t_p, and the crop maturity t_m), the potential aboveground biomass (biom, in tDM ha^{-1}, with1 ha = 0.01 km^2) is computed in the following Eqs. 2.1–C.1. Variables and parameters are summarized in this appendix. For any d in $[t_p, t_m]$, the thermal time (TT) is computed from the daily mean temperature (tas, in °C) by using a reference temperature (T_0):

$$\mathrm{TT}(d) = \mathrm{tas}(d) - T_0 \tag{C.1}$$

$$\mathrm{GDD}(d) = \sum_{i \leq d} \max(0, \mathrm{TT}(i)) \tag{C.2}$$

The number of leaves per plant (n_{leaf}) is computed from GDD and one parameter representing the thermal requirement for the emergence of any leaf ($GDD_{1\mathrm{leaf}}$)):

$$n_{\mathrm{leaf}}(d) = \min(\max_{\mathrm{leaf}}, \mathrm{GDD}(d)/\mathrm{GDD}_{1\mathrm{leaf}}) \tag{C.3}$$

M. Toure, *Agricultural Adaptation to Climate Change in Sub-Saharan Africa*, Advances in African Economic, Social and Political Development,
https://doi.org/10.1007/978-3-031-80716-9

Table C.1 List of variables and parameters in DSSAT model

	Definition	Unit
tas	Daily mean temperature	°C
TT	Thermal time	°C
GDD	Sum of growing degree days	°C
nleaf	Number of leaves per plant	–
rsds	Short-wave downward radiation	$MJm^{-2}d^{-1}$
PARinc	Incoming photosynthetic active radiation	$MJm^{-2}d^{-1}$
APAR	Canopy-absorbed incoming PAR	$MJm^{-2}d^{-1}$
NPPbiom	Net primary productivity dedicated to aboveground biomass	$gDMJm^{-2}d^{-1}$
biom	Aboveground biomass	$gDMJm^{-2}$
NPPgrain	Net primary productivity dedicated to grains	$gDMJm^{-2}d^{-1}$
Grain	Grain biomass (yield)	$gDMJm^{-2}$
T0	Zero of vegetation	°C
GDD1leaf	Sum of growing degree day required for each leaf (phyllochron)	°C
maxnleaf	Maximum number of leaves per plant	–
f	Active fraction of short-wave downward radiation	–
C	C = D k Sleaf dplant with k: coefficient of extinction of radiation in canopy; Sleaf: the specific leaf area of any leaf and dplant: the plant density	–
RUE	Radiation use efficiency	gDM MJ^{-2} (Here, MJ refers to absorbed PAR)
nthresh	Number of leaves from which the formation of reproductive structures starts, the grains form, or the grain filling starts	–
frac	Fraction of NPPbiom going toward the variable grain when nleaf > nthresh	–

Source "Potential yield simulated by global gridded crop models: using a process-based emulator to explain their differences" by Ringeval et al. (2021)

where $\max_{\text{leaf}}$ is the maximum number of leaves. In our model, as soon as one leaf emerges, we assume that it reaches its fully expanded leaf area, called S_{leaf} hereafter). The incoming photosynthetic active radiation (PAR_{inc}) is derived from the short-wave downwelling radiation (rsds in $\text{MJm}^2 d^{-1}$) and its active fraction, f:

$$\text{PAR}_{\text{inc}}(d) = f \times rsds(d) \tag{C.4}$$

The absorbed PAR by the canopy (APAR) is determined by assuming the following exponential function:

$$\text{APAR}(d) = \text{PAR}_{\text{inc}}(d) \times (1 - \exp(-C \times n_{\text{leaf}}(d))) \tag{C.5}$$

where C is a constant. The net primary productivity dedicated to the aboveground biomass (NPP_{biom}) is computed from APAR with a constant radiation use efficiency (RUE in $gDMMJ^{-1}$):

$$NPP_{biom}(d) = RUE \times APAR(d) \tag{C.6}$$

The aboveground biomass corresponds to the sum over time of NPP_{biom}:

$$biom(d) = \sum_{i \le d} (NPP_{biom}(i)) \tag{C.7}$$

The parameter C of Eq. 4.1 can be decomposed in different parameters:

$$C = k \times S_{leaf} \times d_{plant} \tag{C.8}$$

where k is the light extinction coefficient, S_{leaf} is the individual leaf area, and d_{plant} is the plant density. The product of S_{leaf}, d_{plant} and the number of leaves of a given day d (i.e., $n_{leaf}(d)$, corresponds to the leaf area index (LAI) of the same day:

$$LAI(d) = S_{leaf} \times d_{plant} \times n_{leaf}(d) \tag{C.9}$$

To compute the grain biomass at maturity, the day l is defined as follows:

$$n_{leaf(l)} \geq n_{thresh} \tag{C.10}$$

From day l, a fixed fraction (frac) of NPP_{biom} constitutes the net primary productivity dedicated to the variable grain (called NPP_{grain}):

$$\text{if} d \geq l, NPP_{grain}(d) = frac \times NPP_{biom}(d) \tag{C.11}$$

$$f d < 0, NPP_{grain}(d) = 0 \tag{C.12}$$

and finally,

$$grain(d) = \sum_{i \le d} (NPP_{grain}(i)) \tag{C.13}$$

The variable grain(in tDM ha^{-1}) is considered as reproductive structures + grains or grain only. The parameter nthresh is a threshold in the number of leaves from which the formation of reproductive structures starts, in the grains form, or grain filling starts. We focused on biom and grain at maturity, i.e., computed on the last day of the growing season t_m. They are called $biom_{SMM}$ and $grain_{SMM}$. The variable $grain_{SMM}$ is used as the harvested yield.

C.3 Formats of Input and Output Files

The format of model input and output of the DSSAT are text files. Three input files were created to run the DSSAT model using collected data: (i) weather file: Weatherman program in DSSAT and collected weather data, (ii) soil file: SBuild program in DSSAT and soil data, and (iii) experimental data file: XBuild program in DSSAT and crop management data. The following figure shows the format of the DSSAT input file in the Xbuild program which computes the simulated harvested yield of sorghum in 2020. The Xbuild program has seven main sections: treatments, cultivars, fields, planting details, fertilizers, simulation controls, and automatic management (Fig. C.1).

Figures C.2, C.3, C.4, and C.5 display the output format of a DSSAT file. In this simulation output file, there are the following sections: summary of soil and genetic

```
*EXP.DETAILS: NSSN0001SG 2020 SORGHUM YIELD

*GENERAL
@PEOPLE
-99
@ADDRESS
-99
@SITE
-99

*TREATMENTS                        -------------FACTOR LEVELS------------
@N R O C TNAME.................... CU FL SA IC MP MI MF MR MC MT ME MH SM
 1 1 1 0 YY                         1  1  0  0  1  0  1  0  0  0  0  0  1
 2 1 1 0 TA                         1  2  0  0  1  0  1  0  0  0  0  0  1
 3 1 1 0 GR                         1  3  0  0  1  0  1  0  0  0  0  0  1
 4 1 1 0 YY noF                     1  1  0  0  1  0  0  0  0  0  0  0  1
 5 1 1 0 TA noF                     1  2  0  0  1  0  0  0  0  0  0  0  1
 6 1 1 0 GR noF                     1  3  0  0  1  0  0  0  0  0  0  0  1

*CULTIVARS
@C CR INGENO CNAME
 1 SG 990005 CSM 335

*FIELDS
@L ID_FIELD WSTA....  FLSA  FLOB  FLDT  FLDD  FLDS  FLST SLTX  SLDP  ID_SOIL    FLNAME
 1 00000001 YANG       -99   -99   -99   -99   -99   -99 -99    -99 NS15661542 Yang Yang
 2 00000002 TAMB       -99   -99   -99   -99   -99   -99 -99    -99 NS13771367 Tambacounda
 3 00000003 GUIR       -99   -99   -99   -99   -99   -99 -99    -99 NS13041604 Guiragor
@L ...........XCRD ...........YCRD .....ELEV .............AREA .SLEN .FLWR .SLAS FLHST FHDUR
 1             -99             -99       -99               -99   -99   -99   -99   -99   -99
 2             -99             -99       -99               -99   -99   -99   -99   -99   -99
 3             -99             -99       -99               -99   -99   -99   -99   -99   -99

*PLANTING DETAILS
@P PDATE EDATE  PPOP  PPOE  PLME  PLDS  PLRS  PLRD  PLDP  PLWT  PAGE  PENV  PLPH  SPRL                        PLNAME
 1 00167   -99    10   -99     S     R    10   -99     4   -99   -99   -99   -99   -99                        10 seeds/m2,

*FERTILIZERS (INORGANIC)
@F FDATE  FMCD  FACD  FDEP  FAMN  FAMP  FAMK  FAMC  FAMO  FOCD FERNAME
 1 00167 FE005 AP002     4    50   -99   -99   -99   -99   -99 100kg/ha
 1 00207 FE005 AP002     1    50   -99   -99   -99   -99   -99 100kg/ha

*SIMULATION CONTROLS
@N GENERAL     NYERS NREPS START SDATE RSEED SNAME.................... SMODEL
 1 GE              1     1     S 00167  2150 DEFAULT SIMULATION CONTR
@N OPTIONS     WATER NITRO SYMBI PHOSP POTAS DISES  CHEM  TILL   CO2
 1 OP              Y     Y     N     N     N     N     N     N     M
@N METHODS     WTHER INCON LIGHT EVAPO INFIL PHOTO HYDRO NSWIT MESOM MESEV MESOL
 1 ME              M     M     E     R     S     L     R     1     G     R     2
@N MANAGEMENT  PLANT IRRIG FERTI RESID HARVS
 1 MA              R     R     R     R     M
@N OUTPUTS     FNAME OVVEW SUMRY FROPT GROUT CAOUT WAOUT NIOUT MIOUT DIOUT VBOSE CHOUT OPOUT FMOPT
 1 OU              N     Y     Y     1     Y     Y     Y     Y     Y     N     Y     N     Y     A

@  AUTOMATIC MANAGEMENT
@N PLANTING    PFRST PLAST PH2OL PH2OU PH2OD PSTMX PSTMN
 1 PL          20001 20001    40   100    30    40    10
@N IRRIGATION  IMDEP ITHRL ITHRU IROFF IMETH IRAMT IREFF
 1 IR             30    50   100 GS000 IR001    10     1
@N NITROGEN    NMDEP NMTHR NAMNT NCODE NAOFF
 1 NI             30    50    25 FE001 GS000
@N RESIDUES    RIPCN RTIME RIDEP
 1 RE            100     1    20
@N HARVEST     HFRST HLAST HPCNP HPCNR
 1 HA              0 02001   100     0
```

Fig. C.1 DSSAT input file format. *Source* The Decision Support System for Agrotechnology Transfer (DSSAT) version 4.8

```
*SIMULATION OVERVIEW FILE

*DSSAT Cropping System Model Ver. 4.8.0.027 -release  SEP 21, 2023 09:27:56

*RUN   1          : YY                           SGCER048 NSSN0001     1
 MODEL            : SGCER048 - Sorghum
 EXPERIMENT       : NSSN0001 SG 2020 SORGHUM YIELD
 DATA PATH        : C:\DSSAT48\Sorghum\
 TREATMENT  1     : YY                           SGCER048


 CROP             : Sorghum            CULTIVAR : CSM 335          ECOTYPE :IB0001
 STARTING DATE    : JUN 15 2000
 PLANTING DATE    : JUN 15 2000        PLANTS/m2 :    10.0    ROW SPACING :  10.cm
 WEATHER          : YANG   2000
 SOIL             : NS15661542      TEXTURE : LS    -   ISRIC soilgrids + HC27
 SOIL INITIAL C   : DEPTH:200cm EXTR. H2O: 99.8mm  NO3:  0.0kg/ha  NH4:  0.0kg/ha
 WATER BALANCE    : IRRIGATE ON REPORTED DATE(S)
 IRRIGATION       :          0 mm IN      0 APPLICATIONS
 NITROGEN BAL.    : SOIL-N & N-UPTAKE SIMULATION; NO N-FIXATION
 N-FERTILIZER     :       100 kg/ha IN      2 APPLICATIONS
 RESIDUE/MANURE   : INITIAL :       0 kg/ha ;         0 kg/ha IN      0 APPLICATIONS
 ENVIRONM. OPT.   : DAYL=     0.00  SRAD=     0.00  TMAX=     0.00  TMIN=     0.00
                    RAIN=     0.00  CO2 =     0.00  DEW =     0.00  WIND=     0.00
 SIMULATION OPT   : WATER    :Y  NITROGEN:Y  N-FIX:N  PHOSPH :N  PESTS  :N
                    PHOTO    :C  ET      :R  INFIL:S  HYDROL :R  SOM    :G
                    CO2 369ppm   NSWIT   :1  EVAP :R  SOIL   :2  STEMP  :D
 MANAGEMENT OPT   : PLANTING:R   IRRIG   :R  FERT :R  RESIDUE:R  HARVEST:M
                    WEATHER :M   TILLAGE :N

*SUMMARY OF SOIL AND GENETIC INPUT PARAMETERS

   SOIL LOWER UPPER   SAT  EXTR  INIT   ROOT   BULK     pH    NO3    NH4    ORG
  DEPTH LIMIT LIMIT    SW    SW    SW   DIST   DENS                          C
   cm   cm3/cm3    cm3/cm3    cm3/cm3         g/cm3          ugN/g  ugN/g    %
--------------------------------------------------------------------------------
   0-  5 0.046 0.094 0.397 0.048 0.094   1.00   1.32   6.60   0.00   0.00   1.92
   5- 15 0.053 0.102 0.394 0.049 0.102   0.90   1.34   6.60   0.00   0.00   1.63
  15- 30 0.065 0.115 0.390 0.050 0.115   0.80   1.37   6.70   0.00   0.00   1.24
  30- 45 0.074 0.126 0.388 0.052 0.126   0.60   1.42   6.80   0.00   0.00   0.80
  45- 60 0.074 0.126 0.388 0.052 0.126   0.60   1.42   6.80   0.00   0.00   0.80
  60- 80 0.073 0.124 0.388 0.051 0.124   0.00   1.48   7.00   0.00   0.00   0.46
  80-100 0.073 0.124 0.388 0.051 0.124   0.00   1.48   7.00   0.00   0.00   0.46
 100-150 0.064 0.113 0.390 0.049 0.113   0.00   1.54   7.20   0.00   0.00   0.26
 150-200 0.064 0.113 0.390 0.049 0.113   0.00   1.54   7.20   0.00   0.00   0.26

TOT-200   13.3   23.3  77.9   10.0  23.3  <--cm    -  kg/ha-->    0.0     0.0 161348
SOIL ALBEDO     : 0.15        EVAPORATION LIMIT : 4.00          MIN. FACTOR  : 1.00
RUNOFF CURVE # :65.00         DRAINAGE RATE     : 0.75          FERT. FACTOR : 0.90

 Sorghum          CULTIVAR: 990005-CSM 335             ECOTYPE: IB0001
 P1      :413.0  P2O     :12.80  P2R     : 280.0  P5      :640.00
 G1      :  3.0  G2      : 4.00  PHINT   : 49.00  P3      : 152.5  P4     :  81.5
```

Fig. C.2 DSSAT output file format (page 1). *Source* The Decision Support System for Agrotechnology Transfer (DSSAT) version 4.8

```
*SIMULATED CROP AND SOIL STATUS AT MAIN DEVELOPMENT STAGES

 RUN NO.     1      YY

        CROP GROWTH      BIOMASS          LEAF   CROP N      STRESS       STRESS

   DATE  AGE STAGE           kg/ha     LAI   NUM  kg/ha  %    H2O  Nitr Phos1 Phos2
RSTG
 ------ --- ----------     -----   -----  ----   ---  ---  ----  ----  ----  ----
----
 15 JUN   0 Sowing              0   0.00   0.0     0  0.0  0.00  0.00  0.00  0.00
 8
 21 JUN   6 Germinate           0   0.00   0.0     0  0.0  0.00  0.00  0.00  0.00
 9
 25 JUN  10 Emergence           2   0.00   1.1     0  4.8  0.00  0.00  0.00  0.00
 1
 12 JUL  27 End Juveni         95   0.40   9.8     5  5.2  0.00  0.00  0.00  0.00
 2
 28 JUL  43 Panicle In        883   2.12  17.0    31  3.6  0.42  0.00  0.00  0.00
 3
 19 AUG  65 End Lf Gro       3263   5.10  26.3    52  1.6  0.12  0.06  0.00  0.00
 4
 31 AUG  77 Anthesis         4799   4.68  26.3    52  1.1  0.00  0.06  0.00  0.00
 4
  3 SEP  80 End Pan Gr       5181   4.61  26.3    52  1.0  0.00  0.04  0.00  0.00
 5
  4 OCT 111 Maturity         7058   3.33  26.3    99  1.4  0.01  0.00  0.00  0.00
 7
  4 OCT 111 Harvest          7058   3.33  26.3    99  1.4  0.00  0.00  0.00  0.00
 7


*MAIN GROWTH AND DEVELOPMENT VARIABLES

@      VARIABLE                                           SIMULATED      MEASURED
       --------                                           ---------      --------
       Panicle Initiation day (dap)                              43           -99
       Anthesis day (dap)                                        77           -99
       Physiological maturity day (dap)                         111           -99
       Yield at harvest maturity (kg [dm]/ha)                  1999           -99
       Number at maturity (no/m2)                             15610           -99
       Unit wt at maturity (g [dm]/unit)                     0.0128           -99
       Number at maturity (no/unit)                          1561.0           -99
       Tops weight at maturity (kg [dm]/ha)                    7058           -99
       By-product produced (stalk) at maturity (kg[dm]/ha      5059           -99
       Leaf area index, maximum                                5.15           -99
       Harvest index at maturity                              0.283           -99
       Grain N at maturity (kg/ha)                               80           -99
       Tops N at maturity (kg/ha)                                99           -99
       Stem N at maturity (kg/ha)                                18           -99
       Grain N at maturity (%)                                  4.0           -99
       Tops weight at anthesis (kg [dm]/ha)                    4639           -99
       Tops N at anthesis (kg/ha)                                52           -99
```

Fig. C.3 DSSAT output file format (page 2). *Source* The Decision Support System for Agrotechnology Transfer (DSSAT) version 4.8

```
      Leaf number per stem at maturity                          26.31          -99
      Emergence day (dap)                                          10          -99


*ENVIRONMENTAL AND STRESS FACTORS

 |-----Development
Phase------|-------------------------------------Environment-------------------
-----------------|----------------Stress----------------|

|--------------Average-------------|-----Cumulative-----|--------Count of
days--------|           (0=Min, 1=Max Stress)           |
                         Time  Temp  Temp  Temp Solar Photop                 Evapo
   Pot TMIN TMIN TMAX TMAX TMAX RAIN|----Water----|---Nitrogen--|-Phosphorus-|
                         Span   Max   Min  Mean   Rad  [day]    CO2   Rain  Trans
    ET    <    <    >    >    >    >  Photo          Photo         Photo
                         days    ϕC    ϕC    ϕC MJ/m2     hr    ppm     mm     mm
    mm  0ϕC  2ϕC 30ϕC 32ϕC 34ϕC  0mm  synth Growth  synth Growth  synth Growth
--------------------------------------------------------------------------------
------------------------------------------------------------------------------
 Emergence-End Juvenile    17  38.3  24.2  31.3  11.1  12.90  369.3   25.7   14.5
  62.9    0    0   17   17   16    7  0.000  0.000  0.000  0.002  0.000  0.000
 End Juvenil-Panicle Ini   16  36.0  24.4  30.2  10.8  12.80  369.4   15.6   23.5
  49.8    0    0   16   15   12   12  0.286  0.416  0.000  0.000  0.000  0.000
 Panicle Ini-End Lf Grow   22  35.4  24.7  30.0  10.5  12.61  369.4   54.6   61.5
  63.2    0    0   21   21   16   22  0.028  0.123  0.000  0.060  0.000  0.000
 End Lf Grth-Beg Grn Fil   15  34.4  24.6  29.5   9.6  12.37  369.6   77.9   37.7
  37.9    0    0   15   12   10   15  0.000  0.008  0.000  0.060  0.000  0.000
 Grain Filling Phase       30  35.8  24.5  30.1   9.9  12.03  369.6   64.4   82.2
  82.5    0    0   29   28   24   26  0.000  0.013  0.000  0.000  0.000  0.000

 Planting to Harvest      112  36.2  24.4  30.3  10.5  12.51  369.5  240.3  230.0
 342.9    0    0  110  105   89   87  0.046  0.088  0.000  0.020  0.000  0.000
--------------------------------------------------------------------------------
------------------------------------------------------------------------------


*Resource Productivity
 Growing season length: 112 days

 Precipitation during growing season        240.3 mm[rain]
   Dry Matter Productivity                    2.94 kg[DM]/m3[rain]               =
29.4 kg[DM]/ha per mm[rain]
   Yield Productivity                         0.83 kg[grain yield]/m3[rain] =
8.3 kg[yield]/ha per mm[rain]

 Evapotranspiration during growing season  230.0 mm[ET]
   Dry Matter Productivity                    3.07 kg[DM]/m3[ET]                 =
30.7 kg[DM]/ha per mm[ET]
   Yield Productivity                         0.87 kg[grain yield]/m3[ET]    =
8.7 kg[yield]/ha per mm[ET]

 Transpiration during growing season        191.7 mm[EP]
   Dry Matter Productivity                    3.68 kg[DM]/m3[EP]                 =
```

Fig. C.4 DSSAT output file format (page 3). *Source* The Decision Support System for Agrotechnology Transfer (DSSAT) version 4.8

```
36.8 kg[DM]/ha per mm[EP]
   Yield Productivity                          1.04 kg[grain yield]/m3[EP]    =
10.4 kg[yield]/ha per mm[EP]

 N applied during growing season               100. kg[N applied]/ha
   Dry Matter Productivity                     70.6 kg[DM]/kg[N applied]
   Yield Productivity                          20.0 kg[yield]/kg[N applied]

 N uptake during growing season                 114 kg[N uptake]/ha
   Dry Matter Productivity                     61.9 kg[DM]/kg[N uptake]
   Yield Productivity                          17.5 kg[yield]/kg[N uptake]

------------------------------------------------------------------------------------
-----------------------------

                   Sorghum YIELD :      1999 kg/ha     [Dry weight]

************************************************************************************
*****************************
```

Fig. C.5 DSSAT output file format (page 4). *Source* The Decision Support System for Agrotechnology Transfer (DSSAT) version 4.8

input parameters, the simulated crop and soil status at main development stages, the main growth and development variables, environmental and stress factors, resource productivity, and the simulated harvested yield at the bottom of the page expressed in kg/ha.

Reference

Ringeval, B., Müller, C., Pugh, T., Mueller, N., Ciais, P., Folberth, C., Liu, W., Debaeke, P., & Pellerin, S. (2021). Potential yield simulated by global gridded crop models: Using a process-based emulator to explain their differences. *Geoscientific Model Development, 14*, 1639–1656. https://doi.org/10.5194/gmd-14-1639-2021 [60]

Appendix D
Alternatives and Robustness Analysis Supplement Data

D.1 Supplementary Technical Changes Results

See Tables D.1, D.2, D.3, D.4, D.5 and D.6.

Table D.1 Rice simulated production under different cultivars and planting area changes

2050 RCP 2.6	Planting area change	–10%	–5%	0%	5%	10%
	IR64	7.39E+08	7.81E+08	8.22E+08	8.63E+08	9.04E+08
	NE4	9.47E+08	1.00E+09	1.05E+09	1.10E+09	1.16E+09
	DK3	9.99E+08	1.05E+09	1.11E+09	1.17E+09	1.22E+09
2050 RCP 8.5	Planting area change	–10%	–5%	0%	5%	10%
	IR64	6.76E+08	7.14E+08	7.51E+08	7.89E+08	8.26E+08
	NE4	8.40E+08	8.87E+08	9.34E+08	9.80E+08	1.03E+09
	DK3	8.61E+08	9.09E+08	9.57E+08	1.00E+09	1.05E+09
2080 RCP 2.6	Planting area change	–10%	–5%	0%	5%	10%
	IR64	7.13E+08	7.94E+08	8.80E+08	9.70E+08	1.10E+09
	NE4	9.92E+08	1.11E+09	1.22E+09	1.35E+09	1.48E+09
	DK3	7.91E+08	8.81E+08	9.76E+08	1.08E+09	1.18E+09
2080 RCP 8.5	Planting area change	–10%	–5%	0%	5%	10%
	IR64	2.86E+08	3.19E+08	3.53E+08	3.90E+08	4.28E+08
	NE4	5.86E+08	6.53E+08	7.24E+08	7.98E+08	8.76E+08
	DK3	3.88E+08	4.32E+08	4.79E+08	5.28E+08	5.80E+08

Source Author's computations

M. Toure, *Agricultural Adaptation to Climate Change in Sub-Saharan Africa*, Advances in African Economic, Social and Political Development,
https://doi.org/10.1007/978-3-031-80716-9

Table D.2 Maize simulated production under different cultivars and planting area changes

2050 RCP 2.6	Planting area change	−10%	−5%	0%	5%	10%
	Massongo	1.08E+09	1.14E+09	1.20E+09	1.26E+09	1.32E+09
	TZE COMP 4C2	9.87E+08	1.04E+09	1.10E+09	1.15E+09	1.21E+09
	TZB-SR	1.07E+09	1.13E+09	1.19E+09	1.25E+09	1.31E+09
2050 RCP 8.5	Planting area change	−10%	−5%	0%	5%	10%
	Massongo	9.89E+08	1.04E+09	1.10E+09	1.15E+09	1.21E+09
	TZE COMP 4C2	9.22E+08	9.73E+08	1.02E+09	1.08E+09	1.13E+09
	TZB-SR	1.04E+09	1.10E+09	1.16E+09	1.21E+09	1.27E+09
2080 RCP 2.6	Planting area change	−10%	−5%	0%	5%	10%
	Massongo	1.48E+08	1.65E+08	1.83E+08	2.02E+08	2.21E+08
	TZE COMP 4C2	9.03E+08	1.01E+09	1.11E+09	1.23E+09	1.35E+09
	TZB-SR	9.65E+08	1.08E+09	1.19E+09	1.31E+09	1.44E+09
2080 RCP 8.5	Planting area change	−10%	−5%	0%	5%	10%
	Massongo	5.28E+08	5.89E+08	6.52E+08	7.19E+08	7.89E+08
	TZE COMP 4C2	5.93E+08	6.61E+08	7.32E+08	8.07E+08	8.86E+08
	TZB-SR	6.42E+08	7.15E+08	7.93E+08	8.74E+08	9.59E+08

Source Author's computations

Table D.3 Sorghum simulated production under different cultivars and planting area changes

2050 RCP 2.6	Planting area change	−10%	−5%	0%	5%	10%
	CSM335	6.20E+08	6.54E+08	6.89E+08	7.23E+08	7.57E+08
	Improved CSM335	6.54E+08	6.90E+08	7.27E+08	7.63E+08	7.99E+08
	Fadda	6.30E+08	6.65E+08	7.00E+08	7.36E+08	7.71E+08
2050 RCP 8.5	Planting area change	−10%	−5%	0%	5%	10%
	CSM335	6.09E+08	6.43E+08	6.77E+08	7.11E+08	7.44E+08
	Improved CSM335	6.32E+08	6.67E+08	7.03E+08	7.38E+08	7.73E+08
	Fadda	6.06E+08	6.40E+08	6.73E+08	7.07E+08	7.41E+08
2080 RCP 2.6	Planting area change	−10%	−5%	0%	5%	10%
	CSM335	5.44E+08	6.06E+08	6.71E+08	7.40E+08	8.12E+08
	Improved CSM335	5.83E+08	6.50E+08	7.20E+08	7.94E+08	8.71E+08
	Fadda	5.32E+08	5.92E+08	6.56E+08	7.24E+08	7.94E+08
2080 RCP 8.5	Planting area change	−10%	−5%	0%	5%	10%
	CSM335	4.21E+08	4.69E+08	5.20E+08	5.73E+08	6.29E+08
	Improved CSM335	4.45E+08	4.96E+08	5.49E+08	6.05E+08	6.65E+08
	Fadda	4.02E+08	4.48E+08	4.96E+08	5.47E+08	6.00E+08

Source Author's computations

Table D.4 Pearl millet simulated production under different cultivars and planting area changes

2050 RCP 2.6	Planting area change	–10%	–5%	0%	5%	10%
	CIVT*	1.25E+09	1.32E+09	1.39E+09	1.46E+09	1.53E+09
	Zatib*	1.57E+09	1.66E+09	1.74E+09	1.83E+09	1.92E+09
	Heini Kirey*	1.64E+09	1.73E+09	1.82E+09	1.91E+09	2.00E+09
2050 RCP 8.5	Planting area change	–10%	–5%	0%	5%	10%
	CIVT*	1.44E+09	1.52E+09	1.59E+09	1.67E+09	1.75E+09
	Zatib*	1.67E+09	1.76E+09	1.85E+09	1.94E+09	2.04E+09
	Heini Kirey*	1.49E+09	1.57E+09	1.66E+09	1.74E+09	1.82E+09
2080 RCP 2.6	Planting area change	–10%	–5%	0%	5%	10%
	CIVT*	1.04E+09	1.16E+09	1.28E+09	1.41E+09	1.55E+09
	Zatib*	1.28E+09	1.42E+09	1.58E+09	1.74E+09	1.91E+09
	Heini Kirey*	1.42E+09	1.59E+09	1.76E+09	1.94E+09	2.13E+09
2080 RCP 8.5	Planting area change	–10%	–5%	0%	5%	10%
	CIVT*	6.56E+08	7.31E+08	8.10E+08	8.93E+08	9.80E+08
	Zatib*	8.72E+08	9.72E+08	1.08E+09	1.19E+09	1.30E+09
	Heini Kirey*	7.82E+08	8.71E+08	9.65E+08	1.06E+09	1.17E+09

Source Author's computations

Table D.5 Impact of technical changes in simulated maximum and minimum cereal production

	2050 Max production	2050 Min production	2080 Max production	2080 Min production
Maize	1.32E+09	9.22E+08	1.44E+09	5.28E+08
Sorghum	7.99E+08	6.06E+08	8.71E+08	4.02E+08
Pearl millet	2.04E+09	1.25E+09	2.13E+09	6.56E+08
Rice	1.22E+09	6.76E+08	1.48E+09	2.86E+08
Total cereal production with improved cultivars	5.38E+09	3.45E+09	5.92E+09	1.87E+09
Total cereal production with current cultivars	4.52E+09	3.71E+09	3.68E+09	1.89E+09

Source Author's computations

Table D.6 Tabulated aggregate simulated cereal production based on soil profile

	2050				2080			
	Baseline simulated production (kg)		Sandy soil profile simulated production (kg)		Baseline simulated production (kg)		Sandy soil profile simulated production (kg)	
Planting area change	RCP 2.6	RCP 8.5	RCP 2.6	RCP 8.5	RCP 2.6	RCP 8.5	RCP 2.6	RCP 8.5
Rice								
−10%	7.39E+08	6.76E+08	2.12E+08	2.45E+08	7.13E+08	2.86E+08	2.37E+08	8.95E+07
−5%	7.81E+08	7.14E+08	2.23E+08	2.58E+08	7.94E+08	3.19E+08	2.64E+08	9.97E+07
0%	8.22E+08	7.51E+08	2.35E+08	2.72E+08	8.80E+08	3.53E+08	2.92E+08	1.10E+08
5%	8.63E+08	7.89E+08	2.47E+08	2.86E+08	9.70E+08	3.90E+08	3.22E+08	1.22E+08
10%	9.04E+08	8.26E+08	2.59E+08	2.99E+08	1.10E+09	4.28E+08	3.53E+08	1.34E+08
Maize								
−10%	1.10E+09	9.89E+08	7.78E+08	7.12E+08	1.48E+08	5.28E+08	6.80E+08	4.34E+08
−5%	1.10E+09	1.00E+09	8.21E+08	7.52E+08	1.65E+08	5.89E+08	7.58E+08	4.84E+08
0%	1.20E+09	1.10E+09	8.65E+08	7.91E+08	1.83E+08	6.52E+08	8.40E+08	5.36E+08
5%	1.30E+09	1.20E+09	9.08E+08	8.31E+08	2.02E+08	7.19E+08	9.26E+08	5.91E+08
10%	1.30E+09	1.20E+09	9.51E+08	8.70E+08	2.21E+08	7.89E+08	1.00E+09	6.48E+08
Sorghum								
−10%	6.20E+08	6.09E+08	4.66E+08	4.75E+08	5.44E+08	4.21E+08	4.11E+08	3.55E+08
−5%	6.54E+08	6.43E+08	4.92E+08	5.01E+08	6.06E+08	4.69E+08	4.58E+08	3.96E+08
0%	6.89E+08	6.77E+08	5.18E+08	5.28E+08	6.71E+08	5.20E+08	5.07E+08	4.39E+08
5%	7.23E+08	7.11E+08	5.44E+08	5.54E+08	7.40E+08	5.73E+08	5.59E+08	4.84E+08
10%	7.57E+08	7.44E+08	5.70E+08	5.81E+08	8.12E+08	6.29E+08	6.14E+08	5.31E+08
Pearl millet								
−10%	1.25E+09	1.44E+09	5.83E+08	8.51E+08	1.04E+09	6.56E+08	5.12E+08	3.99E+08
−5%	1.32E+09	1.52E+09	6.16E+08	8.99E+08	1.16E+09	7.31E+08	5.70E+08	4.44E+08
0%	1.39E+09	1.59E+09	6.48E+08	9.46E+08	1.28E+09	8.10E+08	6.32E+08	4.92E+08
5%	1.46E+09	1.67E+09	6.80E+08	9.93E+08	1.41E+09	8.93E+08	6.96E+08	5.43E+08
10%	1.53E+09	1.75E+09	7.13E+08	1.00E+09	1.55E+09	9.80E+08	7.64E+08	5.96E+08
All cereals								
−10%	3.71E+09	3.71E+09	2.04E+09	2.28E+09	2.44E+09	1.89E+09	1.84E+09	1.28E+09
−5%	3.85E+09	3.88E+09	2.15E+09	2.41E+09	2.72E+09	2.11E+09	2.05E+09	1.42E+09
0%	4.10E+09	4.12E+09	2.27E+09	2.54E+09	3.01E+09	2.34E+09	2.27E+09	1.58E+09
5%	4.35E+09	4.37E+09	2.38E+09	2.66E+09	3.32E+09	2.57E+09	2.50E+09	1.74E+09
10%	4.49E+09	4.52E+09	2.49E+09	2.79E+09	3.68E+09	2.83E+09	2.75E+09	1.91E+09

Source Author's computations

D.2 Supplement Soil Change Data and Results

See Tables D.7, D.8 and D.9.

Table D.7 Agrimonde-Terra planting area change in 2050 and 2080

		Area of rice (ha)		Area of maize (ha)		Area of sorghum (ha)		Area of pearl millet (ha)	
	Percentage change	2050	2080	2050	2080	2050	2080	2050	2080
Metropolization									
With ultra-processed products	16%	4.73E+05	5.48E+05	3.33E+05	3.86E+05	3.23E+05	3.74E+05	1.20E+06	1.40E+06
With animal products	40%	5.71E+05	7.99E+05	4.02E+05	5.63E+05	3.89E+05	5.45E+05	1.40E+06	2.00E+06
Regionalization									
With agricultural technology A	4.5%	4.26E+05	4.45E+05	3.00E+05	3.13E+05	2.91E+05	3.04E+05	1.10E+06	1.10E+06
With agricultural technology B	11%	4.52E+05	5.02E+05	3.19E+05	3.54E+05	3.09E+05	3.43E+05	1.10E+06	1.30E+06
Healthy									
With agricultural technology C	–4%	3.91E+05	3.76E+05	2.76E+05	2.65E+05	2.67E+05	2.56E+05	9.82E+05	9.43E+05
With agricultural technology D	3%	4.20E+05	4.32E+05	2.96E+05	3.05E+05	2.86E+05	2.95E+05	1.10E+06	1.10E+06
Communities									
With agroecology	18%	4.81E+05	5.67E+05	3.39E+05	4.00E+05	3.28E+05	3.87E+05	1.20E+06	1.40E+06
With collapse	36%	5.54E+05	7.54E+05	3.90E+05	5.31E+05	3.78E+05	5.14E+05	1.40E+06	1.90E+06

Source Author's computations

Table D.8 2050 Agrimonde-Terra simulated cereal production

	2050							
	Simulated production (kg)							
Crops	Rice		Maize		Sorghum		Pearl millet	
Scenarios of planting area change	RCP 2.6	RCP 8.5	RCP 2.6	RCP 8.5	RCP 2.6	RCP 8.5	RCP 2.6	RCP 8.5
Metropolization								
With ultra-processed products	9.53E+08	8.71E+08	1.40E+09	1.30E+09	7.99E+08	7.85E+08	1.60E+09	1.90E+09
With animal products	1.20E+09	1.10E+09	1.70E+09	1.50E+09	9.64E+08	9.47E+08	1.90E+09	2.20E+09
Regionalization								
With agricultural technology A	8.59E+08	7.85E+08	1.30E+09	1.10E+09	7.20E+08	7.07E+08	1.50E+09	1.70E+09
With agricultural technology B	9.12E+08	8.34E+08	1.30E+09	1.20E+09	7.64E+08	7.51E+08	1.50E+09	1.80E+09
Healthy								
With agricultural technology C	7.89E+08	7.21E+08	1.20E+09	1.10E+09	6.61E+08	6.50E+08	1.30E+09	1.50E+09
With agricultural technology D	8.46E+08	7.74E+08	1.20E+09	1.10E+09	7.09E+08	6.97E+08	1.40E+09	1.60E+09
Communities								
With agroecology	9.70E+08	8.86E+08	1.40E+09	1.30E+09	8.13E+08	7.98E+08	1.60E+09	1.90E+09
With collapse	1.10E+09	1.00E+09	1.60E+09	1.50E+09	9.37E+08	9.20E+08	1.90E+09	2.20E+09

Source Author's computations

Table D.9 2080 Agrimonde-Terra simulated cereal production

	2080							
	Simulated production (kg)							
Crops	Rice		Maize		Sorghum		Pearl millet	
Scenarios of planting area change	RCP 2.6	RCP 8.5	RCP 2.6	RCP 8.5	RCP 2.6	RCP 8.5	RCP 2.6	RCP 8.5
Metropolization								
With ultra-processed products	1.20E+09	4.75E+08	2.46E+08	8.78E+08	9.03E+08	6.99E+08	1.70E+09	1.10E+09
With animal products	1.70E+09	6.93E+08	3.58E+08	1.30E+09	1.30E+09	1.00E+09	2.50E+09	1.60E+09
Regionalization								
With agricultural technology A	9.61E+08	3.86E+08	2.00E+08	7.12E+08	7.33E+08	5.67E+08	1.40E+09	8.85E+08
With agricultural technology B	1.10E+09	4.35E+08	2.25E+08	8.04E+08	8.27E+08	6.40E+08	1.60E+09	9.98E+08
Healthy								
With agricultural technology C	8.11E+08	3.26E+08	1.69E+08	6.01E+08	6.18E+08	4.79E+08	1.20E+09	7.47E+08
With agricultural technology D	9.33E+08	3.75E+08	1.94E+08	6.92E+08	7.12E+08	5.51E+08	1.40E+09	8.60E+08
Communities								
With agroecology	1.20E+09	4.92E+08	2.55E+08	9.08E+08	9.34E+08	7.23E+08	1.80E+09	1.10E+09
With collapse	1.60E+09	6.54E+08	3.38E+08	1.20E+09	1.20E+09	9.61E+08	2.40E+09	1.50E+09

Source Author's computations

Appendix E
Resulting Impacts of Climate Change Complementary Data

E.1 Historical and Projected Trend in Child Malnutrition Data

See Fig. E.1.

For **child malnutrition** we utilize a measure of the prevalence of underweight children under five (MAL). The criteria we use for identifying an underweight child is that the child's weight-for-age be more than 2 standard deviations below the median based on National Center for Health Statistics/World Health Organization standards. This measure represents a synthesis of height-for-age (identifying long-term growth faltering or stunting) and weight-for-height (identifying acute growth disturbances or wasting). The large majority of the data, 75%, are from the World Health Organization's Global Database on Child Growth and Malnutrition (WHO, 1997). These data have been subjected to strict quality control standards for inclusion in the database. The criteria for inclusion of surveys in the WHO Global Database are:

- a clearly defined population-based sampling frame, permitting inferences to be drawn about an entire population;
- a probabilistic sampling procedure involving at least 400 children;
- use of appropriate equipment and standard measurement techniques; and
- presentation of data in the form of Z-scores in relation to the NCHS/WHO
- reference population (WHO, 1997).

For **national food availability** we employ countries' daily per capita dietary energy supplies (KCAL). This measure is derived from food balance sheets compiled by the United Nations Food and Agriculture Organization (FAO) from country-level data on the production and trade of food commodities. Given data on seed

M. Toure, *Agricultural Adaptation to Climate Change in Sub-Saharan Africa*, Advances in African Economic, Social and Political Development,
https://doi.org/10.1007/978-3-031-80716-9

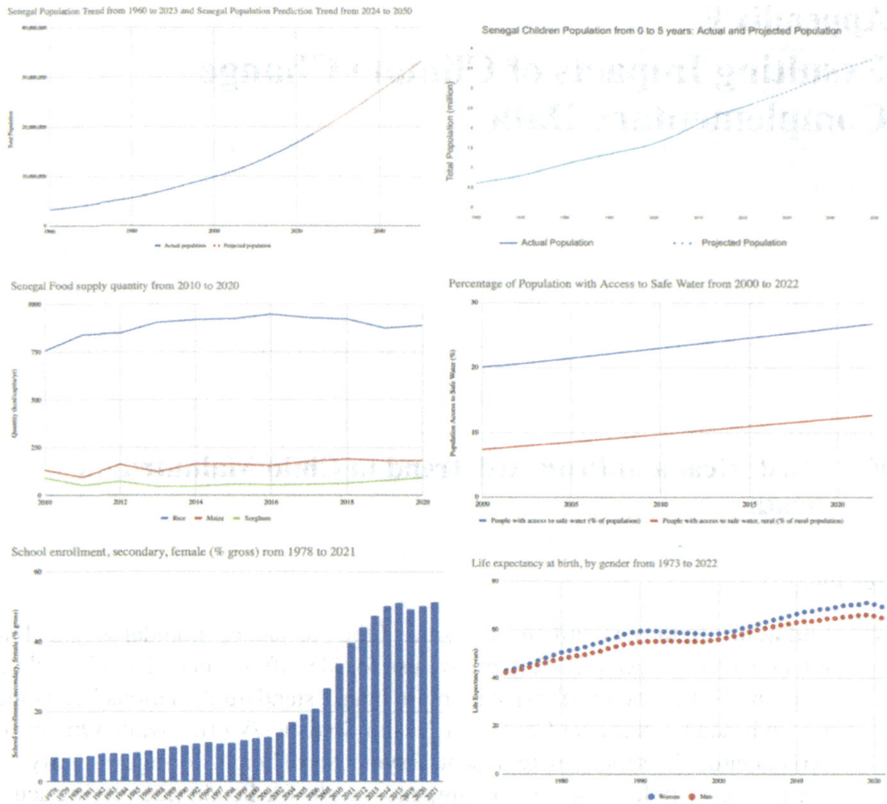

Fig. E.1 Child malnutrition variables. *Source* "Explaining Child Malnutrition in Developing Countries: A Cross-Country Analysis" by Smith and Hadid (2000)

rates, wastage, stock changes, and types of other utilization of food commodities (e.g., animal feed), a supply account is prepared for each commodity in terms of the weight available for human consumption each year. Total energy availability is then estimated by converting the weights of each commodity into energy values and aggregating the energy values across commodities. The aggregate energy supply is then divided by the population size to arrive at per capita KCAL. The data employed were obtained from FAO's FAOSTAT database (FAO, 1998).

For **women's education**, we employ female gross secondary school enrollment rates (SCH) as a proxy measure for female education at all levels up to and including the secondary level. The variable is defined as total female enrollment in secondary education, regardless of age, expressed as a percentage of the population age-group corresponding to national regulations for education at the secondary level. The data are from the United Nations Educational, Scientific and Cultural Organization's UNESCOSTAT database (UNESCO, 1998).

For **women's relative status**, the proxy measure we employ is the ratio of female life expectancy at birth to male life expectancy at birth (LFEXPRAT). Life expectancy at birth is defined as the number of years a newborn infant would live if prevailing patterns of mortality at the time of his or her birth were to stay the same throughout his or her life. The extension of human life reflects the intrinsic value of living as well as being a necessary requirement for carrying out a variety of accomplishments (or 'capabilities') that are generally positively valued by society. It is associated with an enhanced quality of life. Inequalities in this variable favoring males reflect discrimination against females (as infants, children, and adults) and entrenched, long-term gender inequality (Mohiuddin, 1996; Sen, 1998). We believe that the ratio it is a good proxy indicator of the cumulative investments in females relative to males throughout the human life cycle. The source for life expectancy data is the World Development Indicators.

For **access to safe water**, the measure we employ is the percentage of countries' populations with access to safe water (WATER), defined as the population share with reasonable access to an adequate amount of water that is either treated surface water or water that is untreated but uncontaminated water (such as from springs, sanitary wells and protected boreholes). An adequate amount of water is needed to satisfy metabolic, hygienic, and domestic requirements, usually about 20 L per person per day. This measure is used to proxy the broad dimensions of countries' health environments, including access to sanitation and health care (for which insufficient data exist for our use), in that measures of these variables are highly correlated with our measure: countries with high safe water access are likely to have good health environments and services overall. The data are from various issues of UNICEF's State of the World's Children and the World Health Organization.

E.2 Projected Cereal Availability and Consumption Results in 2050

See Tables E.1, E.2 and E.3.

Table E.1 Simulated total cereal availability in Senegal in 2050 (rice, maize, and sorghum)

	Simulated production (kg)		Total cereal produced not exported (kg)		Total cereal imported (kg)		Total cereal available (kg)	
Planting area change	RCP 2.6	RCP 8.5	RCP 2.6	RCP 8.5	RCP 2.6	RCP 8.5	RCP 2.6	RCP 8.5
Rice								
–10%	7.39E+08	6.76E+08	7.39E+08	6.76E+08	7.56E+08	6.92E+08	1.50E+09	1.37E+09
–5%	7.81E+08	7.14E+08	7.81E+08	7.14E+08	7.98E+08	7.30E+08	1.58E+09	1.44E+09
0%	8.22E+08	7.51E+08	8.22E+08	7.51E+08	8.40E+08	7.68E+08	1.66E+09	1.52E+09
5%	8.63E+08	7.89E+08	8.63E+08	7.89E+08	8.83E+08	8.07E+08	1.75E+09	1.60E+09
10%	9.04E+08	8.26E+08	9.04E+08	8.26E+08	9.24E+08	8.45E+08	1.83E+09	1.67E+09
Maize								
–10%	1.10E+09	9.89E+08	1.10E+09	9.89E+08	1.10E+09	9.89E+08	2.20E+09	1.98E+09
–5%	1.10E+09	1.00E+09	1.10E+09	1.00E+09	1.10E+09	1.00E+09	2.20E+09	2.00E+09
0%	1.20E+09	1.10E+09	1.20E+09	1.10E+09	1.20E+09	1.10E+09	2.40E+09	2.20E+09
5%	1.30E+09	1.20E+09	1.30E+09	1.20E+09	1.30E+09	1.20E+09	2.60E+09	2.40E+09
10%	1.30E+09	1.20E+09	1.30E+09	1.20E+09	1.30E+09	1.20E+09	2.60E+09	2.40E+09
Sorghum								
–10%	6.20E+08	6.09E+08	6.20E+08	6.09E+08	6.20E+08	6.09E+08	1.24E+09	1.22E+09
–5%	6.54E+08	6.43E+08	6.54E+08	6.43E+08	6.54E+08	6.43E+08	1.31E+09	1.29E+09
0%	6.89E+08	6.77E+08	6.89E+08	6.77E+08	6.89E+08	6.77E+08	1.38E+09	1.35E+09
5%	7.23E+08	7.11E+08	7.23E+08	7.11E+08	7.23E+08	7.11E+08	1.45E+09	1.42E+09
10%	7.57E+08	7.44E+08	7.57E+08	7.44E+08	7.57E+08	7.44E+08	1.51E+09	1.49E+09

Source Author's computations

Table E.2 Simulated total cereal consumption in Senegal in 2050 (per capita/kcal/year)

	Cereal available quantity (kg)		Cereal per capita food consumption (kg/year)		Cereals per capita calorie consumption availability (kcal/year)	
Percentage change in planting area (%)	RCP 2.6	RCP 8.5	RCP 2.6	RCP 8.5	RCP 2.6	RCP 8.5
–10	4.94E+09	4.56E+09	1.49E+02	1.38E+02	1.14E+03	1.06E+03
–5	5.09E+09	4.73E+09	1.53E+02	1.42E+02	1.18E+03	1.10E+03
0	5.44E+09	5.07E+09	1.64E+02	1.53E+02	1.26E+03	1.18E+03
5	5.79E+09	5.42E+09	1.74E+02	1.63E+02	1.34E+03	1.26E+03
10	5.94E+09	5.56E+09	1.79E+02	1.68E+02	1.38E+03	1.29E+03

Notes With 1 kg = 7.7 kCal. *Source* Author's computations

Table E.3 Simulated net difference in the number of malnourished children (2020–2050)

	Pessimistic scenario		Status quo scenario		Optimistic scenario	
Percentage change in planting area (%)	RCP 2.6	RCP 8.5	RCP 2.6	RCP 8.5	RCP 2.6	RCP 8.5
–10	1.77	3.74	1.51	3.48	1.25	3.22
–5	1.01	2.85	0.74	2.59	0.48	2.33
0	–0.68	1.08	–0.95	0.82	–1.21	0.55
5	–2.27	–0.58	–2.53	–0.84	–2.79	–1.1
10	–2.92	–1.24	–3.18	–1.5	–3.44	–1.76

Source Author's computations

References

Smith, L., & Haddad, L. (2000). Explaining Child Malnutrition in Developing Countries: A cross-country analysis. In *IFPRI Research Report*. International Food Policy Research Institute. [128]

WHO (World Health Organization) (1997). WHO Global Database on Child Growth and Malnutrition. Programme of Nutrition. WHO Document # WHO/NUT/97.4. Geneva.

FAO (Food and Agriculture Organization of the United Nations) (1998). The FAOSTAT database. This database can be accessed through the United Nations Food and Agriculture Organization home page at http://apps.fao.org/cgi-bin/nph-db.pl.

UNESCO (United Nations Educational, Scientific and Cultural Organization) (1998). UNESCO-STAT Database. The database can be accessed through UNESCO's home page at http://unescostat.unesco.org.

Mohiuddin, Y. (1996). Country rankings by the status of women index. In *Paper presented at the 1996 Conference of the International Association for Feminist Economics*.

Sen, A. (1998). Mortality as an indicator of economic success and failure. *Economic Journal, 28*, 1–25.

Appendix F
Climate Finance Supplement Data

F.1 Green Climate Fund

See Table F.1.

F.2 Germany's International Climate Initiative (IKI)

See Tables F.2 and F.3.

M. Toure, *Agricultural Adaptation to Climate Change in Sub-Saharan Africa*, Advances in African Economic, Social and Political Development,
https://doi.org/10.1007/978-3-031-80716-9

Table F.1 List of green fund projects in senegal

Name	Total project value	Tonnes of emissions avoided	Financing	Start date	End (expected) date	Number of co-beneficiary countries
The CATALI.5 °T Initiative	$ 38.8 millions	3.8 millions	Private sector	June 2019	September 2029	16
Inclusive Green Financing Initiative (IGREENFIN I)	$ 188.6 millions	5.6 millions	Public sector	June 2020	April 2022	13
The Africa Integrated Climate Risk Management Programme	$143.4 millions	21.4 millions	Public sector	December 2018	October 2023	7
Global Subnational Climate Fund	$750 millions	77.6 millions	Private sector	August 2017	April 2040	42
Global Subnational Climate Fund—Technical Assistance (TA) Facility	$28 millions	77.6 millions	Private sector	August 2017	April 2040	42
Participation in Energy Access Relief Facility ('EARF')	$60 millions	1.3 millions	Private sector	July 2020	May 2026	9
ASER Solar Rural Electrification Project	$211.2 millions	1.1 millions	Public sector	June 2019	March 2028	1
Promotion of Climate-Friendly Cooking	$60.9 millions	6.5 millions	Public sector	March 2018	September 2025	2
Climate Investor One	$821.5 millions	53.7 millions	Private sector	February 2017	June 2037	19
Transforming Financial Systems for Climate	$693.9 millions	36 millions	Private sector	October 2017	April 2028	17
Building the climate resilience of food insecure smallholder farmers through integrated management of climate risk	$10 millions	–	Public sector	June 2016	January 2024	1
Senegal Integrated Urban Flood Management Project	$75.5 millions	–	Public sector	August 2015	April 2025	1
Increasing the resilience of ecosystems and communities through the restoration of the productive bases of salinized lands	$8.2 millions	–	Public sector	April 2015	February 2026	1

Source Global Green Fund Dashboard, 2023

Table F.2 List of Germany's international initiatives in Senegal (part 1)

Title	Topics	Start date	End date	Amount	Co-beneficiary countries
Blue Solutions—Implementing the CBD Strategic Plan in the field of marine and coastal biodiversity	Knowledge acquisition and improvement of knowledge base	2013-01-01	2024-02-29	€11,150,000.00	15
Catalyzing private financing for climate impact in Africa	Biodiversity finance	2021-01-01	2026-12-31	€20,000,000.00	6
Cities Matter: capacity building in sub-Saharan African megacities for transformational climate change mitigation	Sustainable urban development/urbanization	2017-03-01	2021-12-31	€3,665,375.00	6
Climate Resilience for Critical Sites for Migratory Birds and People along the East Atlantic Flyway	Reduction of loss rate, degradation and fragmentation of ecosystems/areas	2023-05-01	2023-12-31	€443,293.00	2
Ecosystem-based approaches to adaptation: strengthening the evidence and informing policy	Ecosystem-based adaptation (including adapted water and land management)	2015-07-01	2022-12-31	€2,215,355.14	11
Ecosystems Protecting Infrastructure and Communities (EPIC)	Ecosystem-based adaptation (including adapted water and land management)	2012-09-01	2017-08-31	€4,004,645.08	5
Green Banking—Capacity Building for Green Energy and Climate Finance	Financial sector/mobilization of investment in the private sector	2015-11-01	2023-10-31	€4,549,304.22	12
Green Climate Fund—Civil Society Readiness: Focus Africa	Financial sector/mobilization of investment in the private sector	2016-05-18	2023-03-31	€2,896,663.15	9
Institutionalization of CSO engagement in the Adaptation Fund	Financial sector/mobilization of investment in the private sector	2017-05-01	2022-08-31	€1,288,745.00	10
Laying the foundations for long-term low-emission development strategies (LTS)	LCDSs	2021-10-01	2024-05-31	€1,450,000.00	9
Linking Market Mechanisms and Climate Finance in Africa	Carbon market (CDM/JI)/emissions trading	2018-04-01	2023-12-31	€3,949,973.35	2
Mami Wata—Enhancing Marine Management in West Africa through Training and Application	Marine and coastal protected areas	2016-01-01	2022-06-30	€3,763,305.00	17

Source IKI 2023 (https://www.giz.de/en/worldwide/339.html)

Table F.3 List of Germany's international initiatives in senegal (part 2)

Microfinance for ecosystem-based adaptation (MEbA)	Ecosystem-based adaptation (including adapted water and land management)	2012-01-01	2020-12-31	€4,882,952.45	7
National land monitoring and information system for a transparent NDC reporting	REDD+ MRV/REL/Registries	2018-05-01	2022-05-31	€3,150,000.00	21
Non-renewable biomass fraction	Carbon market (CDM/JI)/emissions trading	2011-11-10	2014-06-30	€215,451.00	1
Participation and transparent structures in the post-2015 adaptation architecture	Adaptation strategies/NAP	2010-09-15	2017-01-31	€1,644,343.27	9
Programme for energy efficiency in buildings (PEEB)	Sustainable energy supply/renewable energies/energy efficiency	2017-10-01	2025-12-31	€10,919,196.53	5
Strengthening the capacity of the African Group of Negotiators on Climate Change (AGN) to represent member states in the climate change negotiations and inform the decisions of the African Ministerial Conference on the Environment (AMCEN)	Negotiation support	2016-08-01	2022-12-31	€5,807,723.00	30
Support Program on Scaling up Climate Ambition on Land Use and Agriculture through NDCs and National Adaptation Plans (SCALA)	Adaptation strategies/NAP	2019-11-27	2025-12-31	€20,000,000.00	13
Support to indigenous peoples' and community conserved areas and territories (ICCAs)	Sustainable use of ecosystems/areas	2013-11-01	2024-12-31	€27,000,000.00	35
Supporting PAGE countries on Green Economic Recovery from Covid-19	LCDSs	2020-10-01	2022-12-31	€21,000,000.00	12
Using scientific information databases to support national adaptation plan processes in sub-Saharan Africa	Adaptation strategies/NAP	2016-03-01	2021-09-30	€8,176,237.00	2
West African Alliance on Carbon Markets and Climate Finance	Carbon market (CDM/JI)/emissions trading	2020-08-01	2024-01-31	€2,798,580.00	12

Source IKI 2023 (https://www.giz.de/en/worldwide/339.html)

9783031807152